Newnes Gu...

TV and Video
Technology

Eugene Trundle, MSERT, MRTS, MIST

95

Heinemann   Newnes

1999

Heinemann Newnes
An imprint of Heinemann Professional Publishing Ltd
Halley Court, Jordan Hill, Oxford OX2 8EJ

OXFORD   LONDON   MELBOURNE   AUCKLAND
SINGAPORE   IBADAN   NAIROBI   GABORONE   KINGSTON

First published 1988

British Library Cataloguing in Publication Data
Trundle, Eugene
   Newnes guide to TV and video technology.
   1.   Colour television
   I.   Title   II.   Trundle, Eugene.   Beginner's
   guide to videocassette recorders III. King,
   Gordon J. (Gordon John),   1922-
   Beginner's guide to colour television
   621.388

ISBN 0 434 91986 1

Printed and bound in Great Britain by
Hartnolls Limited, Bodmin, Cornwall

Contents

Preface

This book is a combination of two *Beginner's Guides*: *Colour Television*; and *Videocassette Recorders*. Both books have been overhauled, revised and updated for this single-volume presentation.

Much use is made throughout the book of block diagrams. Since integrated circuits – silicon chips – are now so widespread, much service data are now presented in block diagram form; to explain principles and techniques I have used early 'discrete' circuits, in which the separate functions and circuit elements can be clearly recognised.

This book is aimed at interested laymen, students and technicians and those in allied fields seeking an insight into the technicalities of TV and VTR practice. I have assumed that the reader has a basic knowledge of electrics and mechanics. Although the book addresses itself to domestic products, the line between these and professional gear is now very blurred and ill-defined – the same techniques and even components are used in both these days! Similarly, the various VTR formats are amazingly similar in essence.

For further reading I can recommend my *Television and Video Engineer's Pocket Book* and Steve Beeching's *Videocassette Recorders – A Servicing Guide (3ed)*, both published by Heinemann Newnes. Manufacturers' service manuals and technical training courses are a mine of product-specific information.

My thanks are due to several setmakers for their help and for supplying pictures, individually acknowledged in the captions; and to the Engineering Information Departments of the BBC and IBA.

E. Trundle

1

Basic television

For a reasonable understanding of colour television, it is essential that the basic principles of monochrome TV are known. As we shall see, all colour systems are firmly based on the original 'electronic-image dissection' idea which goes back to EMI in the 1930s, and is merely an extension (albeit an elaborate one) of that system.

Although there are few black and white TVs or systems now left in use, the compatible colour TV system used today by all terrestrial transmitters grew out of the earlier monochrome formats. In the early days it was essential that existing receivers showed a good black and white picture from the new colour transmissions, and the scanning standards, *luminance* signal, and modulation system are the same. What follows is a brief recap of *basic* television as a building block of the colour TV system to be described in later chapters.

Image analysis

Because a picture has two dimensions it is only possible to transmit all the information contained within it in *serial* form, if we are to use but one wire or RF channel to carry the signal. This implies a *dissection* process, and requires a timing element to define the rate of analysis; this timing element must be present at both sending and receiving ends so that the analysis of the image at the sending end, and the simultaneous build-up of the picture at the receiver, occur in synchronism. Thus a television picture may be dissected in any manner, provided that the receiver assembles its picture in precisely the same way; but the path link between sender and viewer must contain *two* distinct information streams: video signal, which is an electrical analogy of the light pattern being sent, and timing signals, or synchronisation pulses, to define the steps in the dissection process. The presence of a timing element suggests that each picture will take a certain period to be built up; how long will depend on how quickly we can serialise the picture *elements*, and this in turn depends on the bandwidth available in the transmission system – more of this later.

Scanning

If we focus the image to be televised on a light-sensitive surface we are ready for the next stage in the dissection process – the division of the pattern into picture elements, or

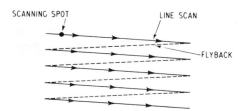

Figure 1.1. The scanning process. Horizontal lines are drawn from left to right of the screen by horizontal deflection, and 'stacked' vertically by the slower-moving vertical deflection field

pixels. Each pixel is rather like the individual dots that go to make up a newspaper photograph in that each can only convey one level of shading. Thus the detail, or definition, in the reproduced picture is proportional to the number of pixels. In 625-line television we have approximately 450 000 pixels, adequate for a 67 cm-diagonal picture, but barely

2

sufficient for much larger screens. These individual pixels are arranged in horizontal lines; there are 625 lines in the British TV system. Figure 1.1 shows how the image is scanned, line by line, to read out in serial form the pattern of light and shade which forms the picture. When half the lines have been traced out the scanning spot has reached the bottom of the picture and traced one *field*. It now flies back to the top of the screen to trace out the rest of the 625 lines in the spaces between those of its first descent. This is known as interlacing, and confers the advantages of a 50 Hz (Hz, Hertz, one cycle per second) flicker rate with the lower scanning speed and lesser bandwidth requirement of a 25 Hz *frame* rate. All TV systems use this 2:1 interlaced field technique; its success depends only on accurate triggering of the field scan.

The camera tube

In practice the image is focused on the internal *target* behind the faceplate of a camera tube, which is usually a variant of the *vidicon* type, illustrated in Figure 1.2. The vidicon has a

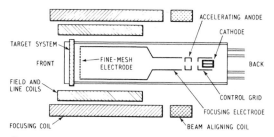

Figure 1.2. Primary elements of photoconductive vidicon camera tube

heated cathode at its back end, in similar style to a picture tube, and this emits a beam of electrons forward towards the target area. A system of electrodes within the tube performs the functions of beam accelerator and electron-lens, so that the beam is brought to pinpoint focus at the target surface.

The scanning action is achieved by suitably-shaped wave-forms in the line- and field-scanning coils surrounding the tube. These magnetically deflect the electron beam on its way along the tube so that the interlaced scanning pattern of Figure 1.1 is traced out on the target. To achieve the very small spot size and precise geometry required of the tiny scanning field, a magnetic focusing system is used in conjunction with the electrostatic lens within the tube, and beam focus optimised by adjusting the DC current in the large cylindrical focus coil. Having set up our image-scanning system, we must now see how the image is read out as an electrical waveform.

Target operation

Figure 1.3 represents the equivalent electrical circuit of the target of a photoconductive camera tube. The target consists of a rectangular plate coated with a photoconductive material

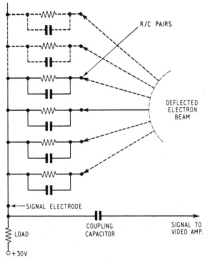

Figure 1.3. Equivalent circuit of photoconductive camera tube, showing the effective RC elements (see text)

4

such as lead oxide whose electrical resistance in the dark is very high, falling to a low level when light is present on its surface. When a complex pattern of light and shade is present, in sharp focus, each picture element will effectively form a parallel RC pair, as in Figure 1.3, and the charge on the capacitor (formed between the front and back layers of the target coating) will be proportional to the light intensity at

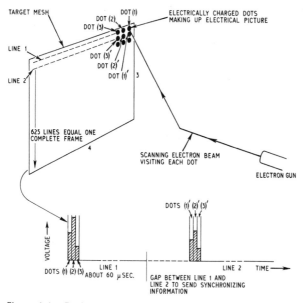

Figure 1.4. Basic principle of scanning at the camera tube

that point. The scanning beam is in effect a conductor, and as it periodically and momentarily picks out each pixel, the capacitor will charge to tube-cathode potential via the electron beam which acts as a conductor. Current for this charging process must come from the target voltage source, so that if we interpose a resistor in series with the positive target potential, we shall develop across it a signal voltage proportional to target current. This is our output signal, an

5

electrical facsimile of the picture element brightness along each line as scanned by the beam. The principle is illustrated in Figure 1.4.

Scanning waveforms

The beam in the camera tube is deflected from right to left of the target (as viewed from within) and its passage takes just $52\,\mu s$ (μs, microsecond, 0.000001 second) during which it passes across the target at a constant velocity. Having reached the left-hand side, the beam flies back to its starting

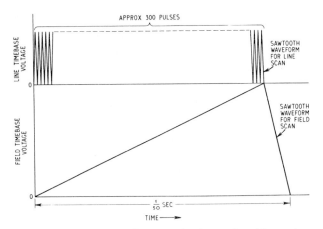

Figure 1.5. Comparison between the time scales of line and field scanning

point on the right, taking about $12\,\mu s$ to do so. While this horizontal scanning process is taking place, the beam is simultaneously being slowly drawn downwards (by the action of the field deflection coil) until it reaches the bottom of the target after 20 ms (ms, millisecond, 0.001 second) whereupon it is suddenly deflected back to the top (field flyback or retrace) to begin the next sweep. Thus the current waveforms in the deflection coils need to be ramp- or sawtooth-shaped with very linear characteristics, and there needs to be 312½ horizontal sweeps (half the total) to each vertical sweep, as in Figure 1.5.

The video signal

The waveform coming from the vidicon target will be similar to that shown in Figure 1.6, consisting of an *analogue* signal representing the picture pattern, with 'blanks' at 64 µs and

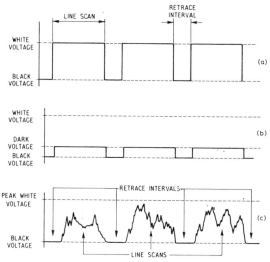

Figure 1.6. Lines of video signal. (a) high brightness. (b) low brightness. (c) picture content

20 ms intervals, during which flyback or *retrace* is taking place. In practice there will be spurious signals during these *blanking* intervals, and these are suppressed in the camera circuit.

Change-coupled devices (CCDs)

An alternative method of televising an image is now commonly used, using a semiconductor pickup device (sometimes called a solid-state image sensor) in place of the traditional thermionic tube. A very small (typically $8 \times 10\,mm$) sensor 'chip' contains a matrix of many thousands of tiny silicon photodiodes, each of which effectively forms a capacitor whose charge is proportional to the brightness of light falling on it. These capacitors are represented by C1, C2, C3 and C4 in Figure 1.7, which portrays the first four pixels in one television line. C1 will have a charge proportional to the light level on the first pixel, and this will appear as a voltage at the output of the first amplifier A1. If all the switches S1 to S3 are now momentarily closed, the charge on C1 will be passed into C2, whose charge will be transferred into C3 and so on, all the way to the last in the line, at the right-hand side of the

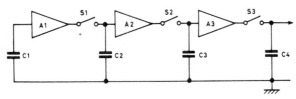

Figure 1.7. The concept of a CCD image sensor. Light patterns are held as charges in the capacitors, one for each pixel

diagram. Thus, by momentarily closing all the switches in synchronism, the brightness level charge for each pixel can be made to 'march' like a column of soldiers to the right where an orderly sequential readout of the light pattern along that line can be picked up, at a rate dependent on the switching frequency. This will form the video signal, with each TV line being read-out in sequence, in similar fashion to a computer's *shift-register*. This is the basis of the operation of a CCD, whose 'scanning' function can be seen to be a *digital* rather than analogue process.

We now have one of our information streams (the video signal), and it's time to insert the second stream (timing pulses) to synchronise the TV receiver. For simplicity, we shall assume that our signal source is a vidicon tube.

8

Synchronisation pulses

The receiver or monitor which we shall use to display the picture has scanning waveform generators too, and these must run in perfect synchronism with those at work in the camera. This will ensure that each picture element picked up from the vidicon target is reproduced in the right place on the display. Plainly, if the camera sees a spot of light in the top right-hand corner of the picture, and the monitor's scanning spot is in the middle of the screen when it reproduces the light, the picture is going to be jumbled up!

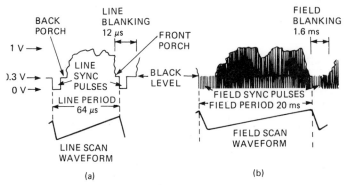

Figure 1.8. The basic analogue TV signal – arrangement of video and sync information and the relationship between signal timing and scanning waveforms

This is prevented by inserting synchronising pulses (sync pulses for short) into the video waveform at regular intervals, and with some distinguishing feature to enable the TV monitor to pick them out. To signal the beginning of a new line scan, we insert a $4.7\,\mu s$ negative-going pulse into each line blanking period, and to initiate field flyback a *series* of similar, closely-spaced pulses are inserted into each field blanking period. These are shown in Figure 1.8, which represents what is called a VBS (video, black-level and sync) or *composite* video signal. Black level is established at $0.3\,V$ ($300\,mV$) from which the signal rises to $1\,V$ for peak white with

9

lesser brightness levels giving correspondingly lower voltage. Each time a sync pulse occurs, the signal voltage drops to zero for its duration. The timing of the field sync-pulse trains is very critical for good interlace in the displayed picture, and the sync pulse generator is carefully designed to achieve this. The short period preceding the line sync pulse is called the *front porch*, and the rather longer ($5.8\,\mu s$) post sync-pulse period is termed the *back porch*. The time spent on porches and sync pulse is known as the *blanking period*, and it's $12\,\mu s$ between lines, and $1.6\,ms$ between fields, as shown in Figure 1.8. The lower section of the diagram indicates the relationship between sync pulses and scanning current for both camera tube and picture tube.

Picture reproduction

We have now obtained a composite video signal, in a form which conveys both video and timing information. Let us now see how it is used to recreate an image on the screen of a picture-tube. For simplicity, we will assume we have a closed-circuit set-up, and that the camera and monitor are linked by a single coaxial cable. Figure 1.9 shows the arrangement. Here we have, at the sending end, a vidicon tube with the necessary lenses, scan coils and power supplies.

Attendant on it is a master sync pulse generator which triggers the sawtooth generators in the camera and also provides a blanking signal for use in the video processing amplifier. A second pair of outputs from its sync-pulse section is taken to an *adder* stage for insertion into the video waveform, and the composite video signal is passed into the transmission cable.

On arrival at the monitor, the signal is first amplified, then passed to the cathode or grid of the picture tube. A second path is to the *sync separator* stage which works on an amplitude-discriminating basis to strip off the sync pulses for application to the timebase generators. They work in just the same fashion as those in the camera to generate sawtooth

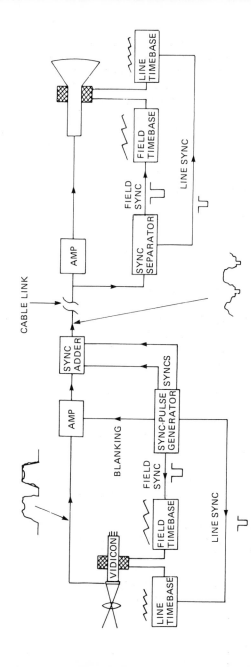

Figure 1.9. A closed-circuit set-up showing derivation and use of the waveforms in Figure 1.8

currents in the scanning coils with which to deflect the scanning beam, but this time in the display tube. Thus we have the two electron beams – one at the sender and one at the receiver – swinging to and fro and up and down in perfect synchronism; and a flow of constantly-changing voltage in the form of the video signal conveying the pattern of light and shade from the vidicon target to the picture-tube screen.

The picture tube

In Chapter 7, we shall study the operation of colour tubes, and as an introduction to these, we need to examine the workings of monochrome tubes – they have much in common! Figure 1.10 shows the basics of a picture tube, which shares many features with the vidicon already described.

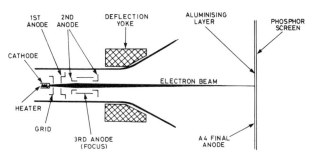

Figure 1.10. The working principle of a monochrome picture tube

The process starts with a heated cathode, from which electrons are 'boiled off' by thermal agitation, to form a *space charge* around the cathode. Depending on the negative potential we choose for the 'grid' (in practice a cylinder surrounding the cathode, with a tiny hole in its otherwise-closed outer face), some of these electrons are attracted away down the tube neck towards the screen by the highly-positive anode cylinders, some of which are shaped and arranged to form an electron-lens whose focal point is the inner surface of the display screen. As the electron beam passes into the 'bowl' of the picture tube it comes under the influence of the scanning coils which deflect the scanning spot, at line and field rate, to trace out a rectangle of light, known as a *raster*, on the inner face of the tube. The screen is *aluminised*, and the thin aluminium layer is held at a potential of several thousand volts with respect to the cathode – this gives an accelerating 'boost' to the electrons in the beam, so that they collide with the phosphor screen surface at very high velocity indeed.

How is the light produced? The inner surface of the faceplate is coated with a continuous layer of *phosphor* which has the property of emitting white light when a high-velocity electron beam impinges on it. The aluminised screen backing reflects the light forward, and forms a barrier to prevent harmful ions burning the screen. The light output is regulated by varying the beam current, and as we have seen, this is the primary function of the 'grid' cylinder. We can arrange a fixed bias on this electrode to set the overall raster brightness, then feed a video signal to it (normally about 80 V peak-to-peak is required) to instantaneously vary the beam current to trace out, and faithfully copy, the brightness of each individual pixel as it is positioned on the screen by the scanning system.

Bandwidth

The scheme outlined so far describes the stages in capturing, relaying and reproducing the picture. We have given little thought as yet to the requirements of the transmission medium, be it a cable, space or a glass fibre!

Mention has already been made of pixels, and the more we have, the faster the rate of change of the video facsimile signal. Much depends on the scanning rate, and this has to be faster that a certain minimum to avoid a disturbing flicker effect in the picture. By using the interlace technique we achieve 50 'flashes' per second, and this is just sufficient to accommodate the human eye's 'time-constant', known as *persistence of vision*. To capture reasonable detail we need all of 625 lines (in fact only about 575 of them are used for the picture itself) and this means that if the 450 000 or so picture elements are each different from their neighbours, the rate of change of the video signal needs to be in the region of 5.5 MHz (MHz, megahertz, one million cycles per second).

Let us look more closely into the reasons for this. If we are televising a picture of a black cat on a white carpet, the rate of change of the video signal will be very low except at the point of the scanning spot's transitions from carpet to cat and back again, each of which will give rise to a sudden *transient*. At the other extreme, if the camera is looking at a pattern of fine vertical lines the video signal will be much 'busier' and contain a great deal of HF energy. In practice the frequencies in the video signal are mostly related to line and field scanning rates, and much of the energy in a video signal is concentrated into 'packets' centred on multiples of line and field frequency. This is an important point, and one to which we shall return.

Modulation

The word modulation, in our context, means the impressing of a signal waveform (usually video, sound or digital pulses) onto a *carrier* wave. The nature of the carrier wave is dependent on the medium to be traversed – space, cable, optical fibre; the distance required to be covered; and to a lesser extent, the nature of the information to be carried. Thus the medium-wave sound broadcast band (MF, medium frequency) is suitable for long-distance broadcasts of rather indifferent-quality sound signals, but quite useless for television; at the other extreme, the SHF (super-high frequency) band is well-suited to a 'beamed' television broadcast service from an orbiting satellite, but one wouldn't expect to find Radio Brighton up there as well! So far as the studio and viewer are concerned, the carrier is irrelevant, because it acts purely as a vehicle on which the wanted signal travels, being discarded at the receiving end once its usefulness has been realised.

Types of modulation

The four types of modulation used in communications to attach information to a carrier wave are (a) amplitude modulation, (b) frequency modulation, (c) phase modulation (which is a variant of *b*), and (d) pulse-code modulation. In ordinary sound broadcasting (a) or (b) is normally adopted, the former at LF, MF and HF, and the latter at VHF (very high frequency, around 100 MHz). In *terrestrial* TV broadcasts, (a) is used for vision and (*b*) for sound. In fact, television uses all four modulation modes, because a form of phase modulation is used for the colouring signal, as we shall see in later chapters, and some modern TV links also make use of PCM.

Amplitude modulation

AM is amongst the earliest forms of modulation to be used, and probably the easiest to understand. It is perhaps surprising that this was pre-dated, however, by PCM (*d* above) in that the very first crude spark transmitters used this mode, conveying the message by simply interrupting the transmitter's operation by means of a morse key! We use AM for the vision signal in terrestrial TV broadcasts; let's see how it is arranged.

For AM, we start with a stable oscillator to generate the carrier wave. For TV broadcasts, the crystal oscillator runs at a sub-multiple of the station frequency, and its output is multiplied up to the required frequency (corresponding to the specified channel number). The information to be sent is made to vary the *amplitude* of the carrier wave, so that in our example of a TV broadcast, the waveform of Figure 1.8 is fed to a *modulator* which may control the *gain* of an RF amplifier which is handling the carrier wave. Thus the video signal is impressed onto the carrier, and the polarity of the video signal is arranged to give *negative modulation*. This means that the tips of the sync pulses give rise to maximum carrier power, and whites in the vision modulating signal raise only about one-fifth of that level. This is illustrated in Figure 1.11 which represents the waveform fed to the UHF broadcasting aerial, and present in the tuner and IF sections of a conventional TV receiver.

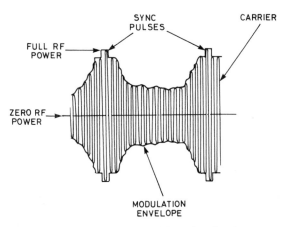

Figure 1.11. AM modulation of the video signal onto an RF carrier for transmission

One of the advantages of AM is the simplicity of the detection or *demodulation* process, which in its original form requires only a rectifier and filter to recover the modulating information.

Frequency modulation

This is the second most popular mode for broadcast use, and is used for high-fidelity (hi-fi) sound transmissions in VHF Band II. Most of these radio transmissions are multiplex-stereo-encoded. So far as we are concerned here, the most significant use of FM modulation is for the TV sound transmissions which accompany 625-line UHF picture broadcasts, and for vision broadcasts from satellites. In this type of modulation the carrier amplitude is kept constant and the frequency of the wave is varied at a *rate* dependent on the frequency of the modulating signal and by an *amount* (deviation) dependent on the strength of the modulating signal (see Figure 1.12). Thus the 'louder' the signal the greater the deviation, while the higher its frequency, the greater the rate at which the carrier frequency is varied either side of its nominal frequency.

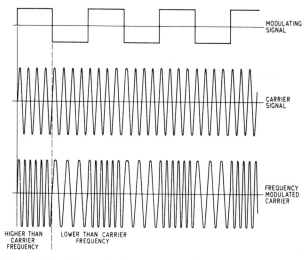

Figure 1.12. Illustrating frequency modulation

Maximum deviation of the TV sound carrier is about ±50 kHz (kHz, kilohertz, one thousand cycles per second) corresponding to maximum modulation depth of this particular system, commonly described as 100 per cent modulation. 100 per cent amplitude modulation is when the troughs of the modulated waveform fall to the zero datum line. This is the absolute physical limit for AM.

Phase modulation

Here the phase of the carrier wave is altered according to the modulating signal. It has much in common with frequency modulation, but the deviation is very small compared with the carrier frequency itself. In a television system, a common form of phase modulation is the transmission of a *reference* signal of constant period alongside a second carrier at the same frequency, but with phase, or timing, variations to convey the required information. This is the basis of the colour encoding system, fully dealt with in later chapters.

Pulse-code modulation (PCM)

An alternative method of modulation is a system of sampling, or quantising, an analogue signal to derive a transmission signal which has only two states, on and off. This is achieved by measuring the instantaneous amplitude of the sound or vision waveform at regular time intervals, short enough to ensure that the highest required frequency is adequately sampled. The amplitude of the signal at each 'sampling spot' is measured and assigned a number which represents the amplitude of the signal at that moment in time. This number can be represented in *binary* (two-state) form, as a series of pulses each of which has a closely defined time slot in a serial transmission. The presence or absence of a pulse in each time slot forms a pattern, unique to each number we may wish to send. The numbers are produced in a continuous string, as it were (known as *serial* transmission). In this form the signal is dispatched, a primary requirement of the transmission medium being sufficient bandwidth to enable it to resolve two recognisable levels (presence of pulse or absence of pulse) at high rates of change. Provided that this requirement is met, the sorts of distortion encountered by the signal on its path do not upset the final reproduction. The second requirement is that some form of *framing* or timing signal accompanies the digital transmission to synchronise the *decoder* at the receiving end.

On receipt, the PCM binary signal passes into a *decoder* which recreates 'numbers' from the pulse train. After processing in a D/A (digital to analogue) converter, the analogue signal is recreated, and its fidelity to the original waveform at the sending end depends on two factors – sampling rate and the number of quantising levels. The sampling rate needs to be at least twice the highest frequency of interest in the *baseband* signal (known as *Nyquist* rate). The number of quantising levels depends very much on the nature of the signal being sampled. For TV an impeccable picture is secured with no more than 256 levels of brightness. Surprisingly, perhaps, an audio signal, if it is to have an acceptable signal-to-noise (S/N) ratio and be free of spurious effects, requires in the region of 1024 quantising levels, though the sampling frequency will be much lower than that for TV pictures. As those who have studied digital electronics will know, this implies an eight-bit 'word' for each pixel of a TV signal ($256 = 2^8$) and a ten-bit word for sound encoding ($1024 = 2^{10}$).

PCM has many advantages over other modulation systems where the transmission path is long, and subject to distortion, reflections and similar hazards for the signal. It is finding increasing applications in television technology, especially in the areas of studio equipment, video tape recording, broadcasters' inter-location links, fibre-optic transmissions and even domestic receiving equipment, though while the transmission system and display device remain in analogue form the benefits of digital processing, particularly in the receiver, are somewhat doubtful.

Sidebands

Whatever type of modulation is used, sidebands are produced. Treated mathematically, the modulated signal can be shown to become a 'group' consisting of a central frequency (the carrier wave) with a pair of symmetrical 'wings' consisting of a number of signals whose displacement from the carrier wave and amplitude depend on the excursions of the modulating signal. This results in a spread of signals across the frequency spectrum, shown for AM modulation in Figure 1.13.

These are the sidebands, and the maximum spread is determined by the highest frequency to be transmitted as modulation. Thus a high-frequency modulating signal will produce sidebands which are located at the limits of the transmission bandwidth. In a vision broadcast these high-frequency modulating signals correspond to the fine detail in the picture, and in a sound broadcast to the upper-treble notes.

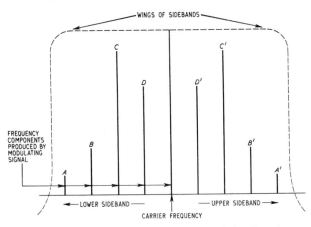

Figure 1.13. Showing the sidebands of a modulated carrier wave

Figure 1.13 shows the central carrier frequency with sidebands for four distinct modulating frequencies simultaneously present. The outer sidebands A and A^1 would correspond to the high frequency (say around 5 MHz) components of the TV picture signal while D/D^1 would arise from lower-definition parts of the baseband signal. C/C^1 and B/B^1 represent intermediate modulating frequencies. Also see page 46.

In FM modulation systems sidebands are also present; their nature is somewhat different, however. Even if only a single pure tone modulating frequency is present, a series of sidebands is produced, and in theory, the 'spread' of the sidebands is infinite. They are spaced on either side of the carrier by multiples of the modulating frequency, although only those relatively close to the carrier are significant in the reception process, depending on ultimate quality. In general an FM transmission system needs more bandwidth than an AM one, as demonstrated by the radio broadcast bands, where a receiving 'window' 20 kHz wide (though it is often less than this in practice, to minimise adjacent channel interference) would afford most adequate reception of MF AM broadcasts, whereas VHF FM reception demands a 'window' of 200 kHz (wider for stereo). For picture transmissions, the comparison is similar – terrestrial AM TV transmitters operate within a bandwidth of 8 MHz, while FM picture signals from satellites demand 27 MHz of spectrum space, although in both cases cited the FM mode affords more 'detail' in the modulating signal, and the bands used for the FM examples are higher in frequency, giving more 'elbow room', so to speak, for the greater sideband spread. We shall return to the subject of sidebands several times in later chapters.

Transmission bandwidth

This band of frequencies has to pass through the various types of equipment, including amplifiers, aerial systems, detectors and so forth in its passage from the studio or programme source to the viewer's home. The bandwidth of such devices will determine just how much of the detail of the original signal gets through. If the bandwidth is too small it will eliminate some of the 'detail' information and also, perhaps, distort the signal in other ways. If, on the other hand, the bandwidth were too great it would allow undue 'noise' and other unwanted signals to enter the system and thus detract from the quality of the vision (and sound). In any

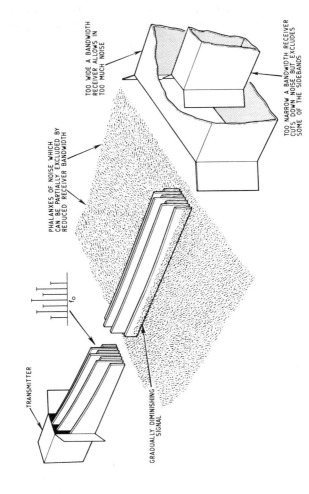

TRANSMITTER

f_o

PHALANXES OF NOISE WHICH
CAN BE PARTIALLY EXCLUDED BY
REDUCED RECEIVER BANDWIDTH

GRADUALLY DIMINISHING
SIGNAL

TOO WIDE A BANDWIDTH
RECEIVER ALLOWS IN
TOO MUCH NOISE

TOO NARROW A BANDWIDTH RECEIVER
CUTS DOWN NOISE BUT EXCLUDES
SOME OF THE SIDEBANDS

Figure 1.14. Model showing the relationship between the sidebands and the bandwidth

22

case the bandwidth spread of any transmission must be limited to the minimum possible spectrum space (rigidly-kept channel widths are specified for all types of transmission) to conserve valuable band space and prevent mutual interference with adjacent channels. A descriptive 'model' illustrating these points is given in Figure 1.14.

2

Light and colour

We know radio and TV broadcas ; as electromagnetic waves whose frequency determines th r wavelength, and various broadcast bands have already be n mentioned in Chapter 1. As we go up in frequency we pas: through the bands allotted to radio transmissions, then te estrial TV broadcast and space communications. Way bey nd these we come into an area where electromagnetic radia on is manifest as heat, and continuing upwards we find infi red radiation, and then a narrow band (between 380×10^6 and 790×10^6 MHz) which represents light energy. Beyond e 'light band' we pass into a region of ultra-violet radiation, rays and then cosmic rays.

Figure 2.1 gives an impressior of the energy distribution curves for red, green and blue lig ts. It can be seen that they

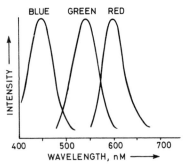

Figure 2.1. Energy distribution curves for red green and blue light

come in the same order as in a rainbow or from a prism. When we see white light it is in fact a mixture of coloured lights of all hues, and the 'splitting' effect of a prism demonstrates this by providing a different refractive index for each light wavelength. If the resulting colour spectrum is passed into a second prism it will be recombined into white light again!

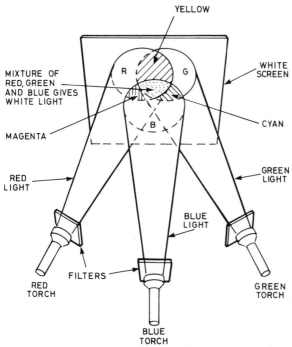

Figure 2.2. White light is produced when red, green and blue lights are caused to overlap on a white screen. The nature of the white light depends on the intensities of the red, green and blue lights. Equal-energy white light is produced when the red, green and blue lights are of equal energy. This is a hypothetical 'white' used in colour studies. White light for colour television is called illuminant D, which simulates direct sunlight and north-sky light mixture

This splitting and recombining process indicates that white light contains all the visible colours, and that by *adding* suitable proportions of coloured light we can create white light. Many combinations of colours may be added to render white, but in television three primary light colours are used: red, green and blue. The principle of *additive* light mixing, as it is called, is shown in Figure 2.2, where light of these three colours is projected onto a screen; in the centre of the display all three colours are present and they add to give a white light.

Colour filtering

Figure 2.2 shows each torch as having a coloured filter over its faceplate. It's important to remember that a filter absorbs all colours except that required, so that the red filter, for instance, will offer little resistance to the low-frequency end of the visible light spectrum, but attenuate green, blue and all other colours. Similarly, a green filter absorbs red, blue etc., allowing only the green component of the white torch beam to pass.

Colour temperature

It is difficult to define just what 'white' light is! The appearance of the white depends entirely on the strength and wavelength of each of the component primaries. Many monochrome picture tubes glow with a bluish, rather cold white, the result of a predominance of high frequencies in the rendered light spectrum; this is because that type of phosphor is more efficient in terms of light output.

Fortunately, an exact definition of the 'whiteness' of a light is available in the form of a *colour-temperature* which relates the nature of the white light to an absolute thermal temperature, that to which a black body must be raised to render the same 'colour' of white. For TV applications this has for many years been standardised at 6500 K, known as Illuminant D, and simulating 'standard daylight'.

Complementary colours

White light may be regarded as a mixture of red, green and blue lights. With these three at our disposal, other colours can be generated by various combinations of two. By removing the blue light we would leave a mixture of red and green lights, which would render yellow, as in Figure 2.2. Yellow is a complementary colour. It is, in fact, complementary to blue since blue was the additive primary which had to be removed from white light to produce it. By similar tokens the complementaries of red and green are cyan and magenta, which means that cyan (akin to turquoise) is produced by the addition of green and blue, and magenta (violet/purple) by the addition of red and blue.

Thus we have the three primaries red, green and blue, and the complementaries cyan, magenta and yellow; all these colours are obtainable – with white light – from the three primary colour *lights*.

It is difficult to visualise the wide range of hues that can be obtained from the three television primaries by changing their relative intensities; but those who view a good-quality colour television picture under correct (low!) ambient lighting conditions will appreciate that almost the full range of natural colours can be reproduced. It is noteworthy that in all the discussions and proposals for TV system improvement and picture enhancement, no change to the primary-colour *additive* mixing scheme has been suggested.

The chromaticity diagram

The range of colours can be conveniently represented by a chromaticity diagram, pictured in Figure 2.3. This is in fact a more elaborate extension of Figure 2.2, showing an elliptical area of various 'whites' at the centre, with the wavelengths of the various colours shown around the periphery. The colours between red and blue have no wavelength references, being 'non-spectral' colours resulting from a mix of components from opposite ends of the visible light spectrum.

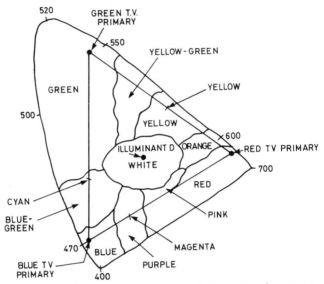

Figure 2.3. The chromaticity diagram. The inner triangle rests on the TV primaries and the area enclosed by the triangle represents the range of colours in a TV display

What we cannot show in a line drawing is the fact that the colours gradually merge into one another around the edges of the diagram. The three TV primaries are shown, and linked as a triangle – it can be seen that for any primary, its complement appears about half-way along the opposite side of the triangle.

Luminance, hue and saturation

Basic television, as we saw in Chapter 1, is concerned solely with brightness variations, and the video waveform conveys information which describes only the variations of light and shade in the picture, as does a black-and-white photograph. As such it works only at a point in the middle of our chromaticity diagram, and so far as the viewer is concerned, the precise point depends on the makeup of the phosphor used in his picture tube! The monochrome signal with which we have been dealing so far is called a *luminance* signal.

To describe a colour, two more characteristics must be defined, and these are hue (colour) and saturation ('strength' of colour). Let's look at each in turn. The hue is set by the dominant wavelength of the light radiation, be it red, blue, cyan or whatever, and corresponds to points around the edges of our chromaticity diagram. Figure 2.3, however, encloses an *area* and real colours can occupy any point within that area. Very few colours in nature are fully *saturated* (which would set them on the periphery of the diagram); most fall within it, and this is where the second parameter, saturation, becomes relevant.

Saturation describes the amount by which the colour is diluted by white light, which in terms of the colour triangle in Figure 2.3, tells us how far from the white centre is the colour in question. If one can imagine a pointer whose origin is Illuminant D in the centre, and whose head pointed to the colour of interest, its length would represent saturation, and its *angle* would describe hue. Comparing the healthy pink of a child's face and the bright red of a London 'bus, it's not hard to see that both have the same hue – red – but the saturation levels are very different, the colour coming from the child's face corresponding to a point near the centre of the chromaticity diagram. This can be graphically illustrated on some colour TV receivers by lowering brightness and contrast and grossly over-advancing the colour level – the face will be reproduced in as strong a red as the 'bus in a normal picture!

The human eye

The phenomenon of persistence of vision has already been touched upon. The eye tends to retain an image for around 80 milliseconds after it has disappeared, and advantage is taken of this in both television and cinematography, where a series of still frames create the illusion of a continuous moving picture. Other important characteristics of the eye are its relative insensitivity to *coloured* detail in a scene, and its failure to respond equally to all colours. Its greatest sensitivity is in the region of yellow/green, with lesser response in the areas of red and particularly blue.

Contrast ratio

In nature the range of brightness levels is infinite, from the brilliance of the sun to the total darkness of an enclosed cave. When an image is reproduced in photographic or TV form, the difference between the brightest and darkest parts of the picture is greatly reduced, and so far as TV reproduction is concerned, much depends on the level of ambient light. The darkest parts of the picture can be no blacker than the unenergised phosphor screen, and even with no external lighting source to illuminate the screen, a degree of reflection of the highlight picture components is present from the viewing area, and indeed the viewers faces! Because the maximum brightness from a picture tube is limited, a contrast ratio of about 50:1 is usual in normal conditions. With a large light-absorbing auditorium and much higher light energy available, the cinema does much better than this.

Gamma correction

In the previous chapter we briefly examined the camera tube and picture tube in their roles of pickup and display devices. In terms of light input to voltage output, the vidicon tube is quite linear so that equal increments of brightness in the televised scene will give rise to equal steps of voltage at the target output point. Unfortunately, the picture tube is not linear in its operation. If we apply an equal-increment staircase waveform to the electrical input of the picture tube, the light output will not go up in corresponding steps; at high brightness levels the graduations will be emphasised or stretched, whereas low-key steps will be compressed. This means that the video signal needs to pass through a compensating circuit, with a gain/level characteristic equal and opposite to that of the tube. It would be expensive to provide such a gamma-correcting amplifier in every receiver, so the process is carried out at the transmitting end, to 'pre-distort' the luminance signal and cancel out the display-tube non-linearity.

3

Reading and writing in three colours

If we bring together the themes of the last two chapters we are well on the way to realising a form of colour television system. Because all the colours in the scene to be televised can be analysed in terms of the three television primaries, we can assemble a colour TV outfit by triplicating the 'basic'

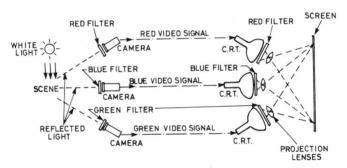

Figure 3.1. Elementary colour TV system working on the 'simultaneous' principle

television system and assigning one primary colour to each of the three. Figure 3.1 shows the set-up. Three identical cameras are used, each with an appropriate filter in front of its pickup tube. The three transmission cables carry video signals which correspond to the three primaries: R, G, and B for short. The monitors also have colour filters fitted, and their pictures are superimposed either by projection onto a common screen, or by a series of dichroic mirrors.

The colour picture reproduced by this set-up would be very good indeed once perfect superimposition, or *register*, of the three images had been achieved at both ends. As a system, however, its shortcomings are immediately obvious: three cameras, three separate transmission channels using (taking normal 625/50 parameters) a total of 16 MHz bandwidth, three identical receivers and goodness knows what in the form of filters and optics at each end! Not only this, but the system is quite incompatible with any monochrome (black-and-white mode) apparatus at either end.

The sequential approach

We know that basic television is no more than an advancing series of still frames, following each other quickly enough to prevent visible flicker. An early idea was to extend this technique to colour reproduction by mechanically changing the colour filter over the pickup tube at field or frame rate in the sequence of RGBRGB . . . so that 'snapshots' of each of the primary-colour components of the televised scene are sent out in quick succession. The single transmission system would now require only one TV receiver feeding a single picture tube equipped with three sequentially-switched colour filters working in synchronism with those at the camera.

The problem here is one of flicker. Because the repetition rate of each primary colour is now only one-third of the (just adequate) TV field repetition frequency, coloured objects flicker alarmingly; even if a form of 'storage' were devised to overcome this, moving coloured objects would traverse the screen in a series of hops. To overcome the problem a threefold increase in field-scanning rate would be required, calling in turn for three times the signal bandwidth throughout the system. Plainly, a radically different method of sending colour pictures was required, and the fact that much of the information in the R, G and B signals is common to all three (oscilloscope examination of these waveforms on a normal picture defies anyone to tell them apart!) is a key factor in the solution devised.

Compatible colour television

Before we explore the practicalities of the colour TV system as it exists today, it is useful to provide a brief overview of how the problems described above are overcome in a modern system. In place of the three cameras we have a single camera which may contain between one and four pickup devices, with the necessary colour separation carried out by optical filters near or within the pickup tubes or sensors. The output from the camera is contained in one video transmission channel no more that 5.5 MHz wide by an encoding process, designed to render a signal which is compatible with monochrome receivers.

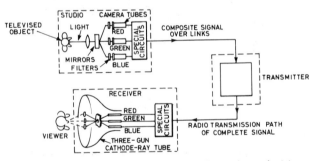

Figure 3.2.   Elementary concept of a complete colour television system

At the receiving end the encoded CVBS (colour, video, blanking and syncs) signal is handled in a single receiving channel identical to that of a monochrome set, so that only one tuner and IF amplifier are needed. The *luminance* and *chrominance* components of the signal are reassembled into RGB signals in a *decoder* and presented to a colour display

34

device (usually a *shadowmask* picture tube) which is capable of reproducing all three primary-colour images on a single screen. This is summed up in Figure 3.2, and much of the first half of this book will be devoted to explaining the processes outlined above!

Three-tube camera

Probably the simplest type of camera to understand is the three-tube type, so we shall adopt this as a basis for an account of the operation of the first link in the chain. A multi-tube camera has only one lens, of course, so the incoming light has not only to be colour-filtered, but 'split' and distributed between the three camera tubes. This is

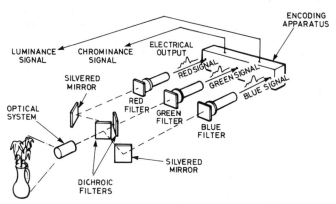

Figure 3.3. Basic elements of the front end of a colour-transmitting system

achieved by a system of dichroic filters and silvered mirrors as outlined in Figure 3.3. The dichroic filter is a precision-engineered optical device capable of reflecting one colour while appearing transparent to others. In this way each pickup tube is presented with an image on its active surface composed of one primary-colour component of the picture to be televised.

The output from each camera tube is thus the electrical fascimile of the amount of each primary colour present in the scene. These outputs are now ready for processing to produce the compatible CVBS signal referred to earlier.

Deriving a luminance signal

It is essential that any colour system produces a signal which is recognisable to a monochrome receiver or monitor, so that the viewer has the choice between a colour or black-and-white set. This requires that the colour transmission conforms to all the specifications of the 'basic' TV system, so we must send primarily a luminance signal to convey brightness information, then add to it as best we can further information streams, used only in colour receivers, to describe the colour. This meets the compatibility requirement by permitting the use of monochrome picture sources at the sending end, and the choice of receiver at the receiving end. It has the incidental advantage of providing black-and-white reproduction when the receiver's decoder breaks down! Our first task, then, is to derive a luminance signal from the RGB outputs of the three pickup tubes.

The luminance signal is produced by adding the red, green and blue signals from the camera tubes in the respective proportions of 30, 59 and 11 per cent, making a total of 100 per cent. This is best seen in simple algebraic terms such that the luminance signal, denoted Y in colour television, is $Y = 0.3R + 0.59G + 0.11B$. Here R, G and B correspond to the primary colours of red, green and blue. If each primary-colour signal from the corresponding camera tube is initially adjusted on a pure peak white scene for 1 V, then by taking 30 per cent of the signal from the red tube, 59 per cent of the signal from the green tube and 11 per cent of the signal from the blue tube and adding them all together we would obtain a total of 1 V Y signal. The tubes, of course, are not themselves coloured, any more than the green channels, blue electron beams and red guns we shall discuss later in the book! The 'colour labels' so applied merely refer to the circuit or channel in which the component in question is operating.

It may be wondered why equal proportions of R, G and B are not combined to produce the luminance signal. In fact the proportions chosen correspond to the sensitivities to each colour of the human eye, discussed in Chapter 2, and this ensures that the luminance display (the only one available to a monochrome set) appears to the viewer to be *panchromatic*.

This luminance signal, then, corresponds closely with that termed the video signal in a monochrome system. In fact, from first principles a monochrome (single-tube) camera adjusted to yield 1 V signal from a pure white 'object card' would produce (under the same lighting conditions) 0.3 V output from a saturated red card, 0.59 V from a saturated green card, and 0.11 V from a saturated blue card – the same proportions as given from the pickup tubes of a colour camera scanning a pure white card, the three adding to produce 'unity' white signal. In practice the idea works well, though the necessary gamma-correction process introduces mild brightness errors on monochrome receivers in areas of the picture where heavily-saturated colours are present.

Single-tube cameras

In response to a healthy demand from users of portable VTR equipment, technology was developed for the practical realisation of a single-tube colour pickup device. The single-tube camera has a lot to offer in terms of economy and simplicity since it obviates the need for expensive optical hardware, and poses no registration problems. These advantages have a trade-off in the realm of picture quality, however, but this is less significant when (as is usually the case) such cameras are used in conjunction with a domestic-format VTR machine whose inherent shortcomings mask those of the camera.

All single-tube cameras to date depend for their operation on a special target system, in which a matrix is present to give a colour-discriminating property. The matrix is in the form of an optical filter, and some tubes also have a matrix target. A good example of the latter type is the tri-electrode tube developed by Hitachi, and shown in Figure 3.4. Behind the glass faceplate of what is basically an ordinary vidicon tube is a striped optical filter consisting of vertical strips in the order RGBRGB etc. Behind this are three grilles, one for each primary colour; their vertical bars are physically aligned with the corresponding optical colour stripes. The grilles are in

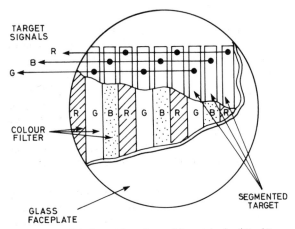

Figure 3.4. Tri-electrode colour pickup tube by Hitachi – operation is explained in the text

fact segmented targets which are scanned in the normal way by the vidicon beam, and the outputs rendered by these three stripe targets are RGB video signals corresponding to the outputs from the separate tubes in the conventional camera. A Y signal is obtained by adding correct proportions of R, G and B as before.

CCD colour cameras

The CCD pickup device described in Chapter 1 is now widely used in colour cameras. It has the advantages over thermionic devices of small size, and physical and optical ruggedness. As with conventional tubes, either three, two or one device can be used for full colour pickup, though single-and two-sensor cameras need some form of filter matrix along the lines of the stripe-filter vidicons. The early problems of low definition, halation and 'vignetting' in CCD sensors have now been overcome, and devices with up to 440 000 pixels are currently fitted to semi-professional and 'home' cameras.

Colour-difference signals

We are now aware that a coloured scene possesses three important characteristics. One is the brightness of any part of it, already discussed as luminance; two is the hue of any part, which is the actual colour; and three is the saturation of the colour, that is the depth of colour. Any hue can vary between very pale and very deep, and the amount of the colour is basically its saturation. For example, a laurel leaf represents a highly saturated green, while the pastel green of my blotting paper is the same hue but less saturated, which means that on a TV screen its image would have a greater contribution of white.

When a camera tube is scanning the red in a scene, as an example, it receives information on the luminance, hue and saturation because all three are obviously present in any colour. However, remember that the luminance signal is processed separately by proportioned addition of the primary-colour signals and transmitted effectively in 'isolation' so that it can be used by monochrome receivers. This, then, means that an additional signal has to be added to and transmitted with the luminance signal so that colour receivers will obtain the extra information they require to describe the hue and saturation of the individual pixels. This auxiliary signal may be regarded as the 'colouring' agent; it is called the *chroma* (short for chrominance) signal.

It is formed initially on a *subcarrier* which is then suppressed at the transmitter and recreated in the receiver, as we shall see later. The subcarrier is modulated in a special way by *colour-difference* signals, of which there are three but only two of them need to be transmitted.

The three colour-difference signals are red minus the luminance signal, green minus the luminance signal, and blue minus the luminance signal. By red, green and blue is meant the primary-colour signals delivered by the colour camera, while the luminance signal is the Y component as defined by the expression given earlier.

Thus, in simple algebraic terms the three colour-difference signals are: $R - Y$, $G - Y$ and $B - Y$.

It is not necessary to transmit all three colour-difference signals. If we send Y, $R - Y$ and $B - Y$, it is easy to recreate $G - Y$ at the receiver. Let's see how this comes about. At the receiver we have available the Y signal, and information on how *different* is the R signal (in $R - Y$) and how *different* is the B signal (in $B - Y$). When these *combined* differences are taken into account, any outstanding balance (or debit!) represents the G difference, so that a $G - Y$ signal can be derived by adding suitable proportions of *negative* $R - Y$ and $B - Y$ signals in the receiver's decoder.

The reason behind the choice of $R - Y$ and $B - Y$ signals for transmission is very simple. Although any two difference signals could be sent, we have seen that the Y signal draws more heavily on the green primary-colour component of the televised image than either of the other two (Y contains 0.59G), so it follows that the difference between G and Y is less than that for the others. This relatively small $G - Y$ signal would be more vulnerable to noise in the transmission system than the larger $R - Y$ and $B - Y$ signals.

It is interesting to observe that the colour-difference signals fall to zero amplitude when the televised scene is devoid of colour – that is, when greys and whites are being transmitted. This is not really surprising when we remember that the chroma signal is composed of colour-difference components – in a *monochrome* transmission there is no difference between the primary-colour signals and the luminance signal! Monochrome programmes are becoming rare now, consisting mainly of archive material and old feature films.

We have seen that the red, green and blue signals from the tubes of a colour television camera can be conveniently tailored to 'unity' (1 V) on a pure peak white input. Thus $R = 1$, $G = 1$, and $B = 1$. Y, we have seen, is equal to $0.3R + 0.59G + 0.11B$, which means that on a pure peak white input we have $0.3(1) + 0.59(1) + 0.11(1)$, which equals 1. Clearly, then, from this we get $R - Y = 1 - 1 = 0$ and $B - Y = 1 - 1 = 0$. The same conditions exist on greys when the red, green and blue signals from the tubes are less than 'unity' but still equal. On colour scenes, of course, the RGB signals are not equal and so colour-difference signals arise, and only when this happens is chroma signal produced.

It is possible, of course, to calculate both the luminance signal and the colour-difference signals from the colour scanned by the colour camera, remembering that anything below full saturation means that white is added to the predominant hue in terms of the three primary colours in the proportions required for the luminance signal. Thus, while the R, G and B signals become unequal when the camera is scanning a coloured scene, the Y signal still retains the proportions of 0.3, 0.59 and 0.11 of the R, G and B signals respectively.

For example, purple of below full saturation comprises a mixture of red and blue with a little green signal too, so that the Y proportions of the red, green and blue provide the 'white' which reduces the saturation. Thus we may have $R = 0.6$, $G = 0.1$ and $B = 0.5$, meaning that the luminance signal Y is equal to $0.3(0.6) + 0.59(0.1) + 0.11(0.5)$, or $0.18 + 0.059 + 0.055 = 0.294$. Using this for Y, then, $R - Y$ is $0.6 - 0.294 = 0.306$ and $B - Y$ $0.5 - 0.294 = 0.206$. When the Y signal is of a greater voltage than the primary colour components of the colour-difference signal, the colour-difference signal as a whole then assumes a *negative* value as, of course, would be expected. The colour-difference signals, therefore, can swing from zero to a maximum in both the positive and negative directions. A little thought will confirm that if we have a white raster displayed, and wish to change it to blue or yellow, the means of doing so is to provide a positive or negative blue colour-difference signal. For saturated blue, the red and green colour-difference signals would also operate to turn off their respective colours.

Colour-difference signal bandwidth

The colour-difference signals are used, as we have seen, to *add* colour information to the basic black-and-white picture. Because the human eye cannot resolve fine detail in colour, there is little point in transmitting high-definition colour-difference signals, which would be difficult to accommodate in the signal channel, and wasted on arrival! The luminance signal carries all the fine detail in the picture, then, and the colouring signal is 'overlaid' on the display in much coarser form. In practice the illusion is well-nigh perfect, and subjective viewing of the combination of high-definition luminance and rather 'woolly' chrominance is perfectly satisfactory. In the American colour system they go a step further, and transmit even lower definition in their 'Q' signal, corresponding to blue shades, where research shows that the eye is least able to discern detail. Reproduced colour pictures from domestic videocassette recorders offer even poorer chrominance definition, but still with (just) satisfactory results.

In the UK broadcast colour system, then, we limit the bandwidth of each colour-difference signal to about 1.2 MHz by means of electrical bandstop filters.

The encoder

The three primary-colour signals from the three camera tubes are communicated to the input of the encoder, and this processes them ready for transmission. The process can be likened to the sending of an important message in code. If a code book is used at the dispatch end, the letter can be reduced to a shortened form and sent to the recipient who, using a similar code book, can decipher it and thereby recreate the original message.

The three primary-colour signals are first added together in the proportions required for the luminance or Y signal, as already explained. The resulting Y signal is then separately subtracted from the red and blue primary colour signals to give the two colour-difference signals R − Y and B − Y. These two signals are amplitude-modulated in quadrature (see anon) upon a subcarrier which is subsequently suppressed so that only the sidebands of the V and U signals remain.

In monochrome television (now largely confined to industrial and special-purpose applications) only the luminance signal and the sync and black level are modulated onto the broadcast carrier wave.

With the need to send a chroma signal in addition to the basic VBS (Video, Blanking and Syncs) signal, coupled with the requirement to keep the combined signal within the channel bandwidth normally occupied by a monochrome signal, the method of fitting together all the components of a signal for compatible colour television is necessarily more complex. As already outlined, the scheme utilises a colour subcarrier of a much lower frequency than the main carrier wave. The latter, in fact, may be in the hundreds of MHz or the GHz (GHz, gigahertz, one thousand million cycles per second) range, whereas the former is a few MHz only. It is within the luminance bandwidth range in fact, the actual frequency being geared to the line and field timebase repetition frequencies.

We shall be seeing later that to get the chroma signal to integrate neatly with the luminance signal, the frequency of the subcarrier must be related to line- and field-scanning frequencies. The extra 'colouring' information is then squeezed into discrete intervals of low energy between the luminance sidebands in the overall frequency spectrum. It is by this technique that the extra chroma information can be carried in an ordinary 625-line television channel with the least mutual interference, especially to black-and-white pictures produced by monochrome receivers working from a colour signal.

This clever technique, of course, calls for a comprehensive type of sync-pulse generator, for its master oscillator must be correctly related to the frequency of the subcarrier generator. However, in practice the master oscillator generates subcarrier frequency, which is then divided down by counter circuits to derive line and field synchronisation pulses. This ensures that the relationship

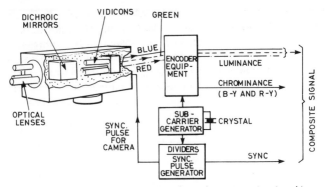

Figure 3.5. Showing in basic terms how the composite signal is developed. The subcarrier generator provides a reference for both sync and colour signals, and the composite signal developed contains all information necessary for the recreation of a colour picture

between the three is correct. Figure 3.5 outlines the processes described thus far.

Composition of the chroma signal

Having discussed the derivation of luminance and colour-difference signals, and seen how all the characteristics of a full colour picture can be carried in them, it's time now to explore the way in which they are combined to make up a CVBS (Chroma, Video, Blanking, Syncs) signal. We have already referred to the process of quadrature modulation onto a subcarrier – just what does this mean?

The basics of amplitude modulation have been discussed in Chapter 1. The modulation signal of lowish frequency modulates the carrier wave of higher frequency by causing its amplitude to vary in sympathy with the modulating signal. Figure 3.6 illustrates the effect. Here a single sinewave signal is shown modulating a higher-frequency carrier wave. This modulation is quite easy to understand from an elementary viewpoint, and it can be achieved by the use of simple circuitry. For example, the modulating signal can be applied to a transistor so that it effectively alters the supply voltage on the collector. Thus when the carrier wave is applied to the base, the amplitude of the output signal alters to the pattern of the modulation waveform. When a pure sine wave is the

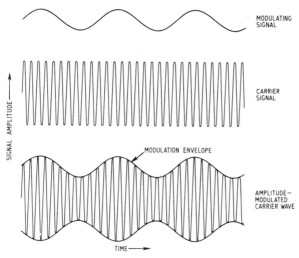

Figure 3.6. The amplitude modulated waveform is produced by the carrier signal being modulated with the modulating signal. Note that the waveform inside the modulation envelope results from the addition of the carrier wave (fc), the upper sideband (fc + fm) and the lower sideband (fc − fm), where fm is the modulation frequency

modulation signal, therefore, the carrier-wave amplitude varies in a like pattern and the carrier wave is then said to possess a modulation *envelope*, as shown at the bottom of Figure 3.6.

We have already seen that any type of modulation gives rise to sideband signals, and for amplitude modulation there is an upper and a lower sideband for each modulation frequency. If the carrier wave is, say, 10 kHz and the modulating signal a pure 1 kHz sine wave (e.g. a single modulation frequency) then the upper sideband will be $10 + 1 = 11$ kHz, while the lower sideband will be $10 - 1 = 9$ kHz. This simple arithmetic follows for all single modulation frequencies. Thus the modulator delivers three signal components, the carrier wave, the upper sideband and the lower sideband. The information due to modulation is effectively present in the sidebands, so it is possible to suppress the carrier after modulation for transmission, though for demodulation at the receiver the carrier wave will need to be somehow reintroduced very accurately.

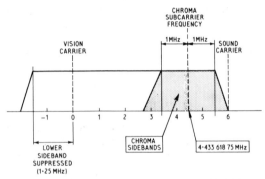

Figure 3.7. Spectrum of 625-line PAL signal in UK television channel

So much, then, for the amplitude modulation of one set of information, but what about the modulation of two sets of information which convey the $R - Y$ and $B - Y$ colour-difference signals? This is where the quadrature amplitude modulation of the subcarrier comes in. In the UK system of colour television the subcarrier frequency is accurately controlled at 4.43361875 MHz. This is usually referred to roughly as 4.43 MHz. The numerous decimal places (the last of which defines the frequency to one-hundredth of one cycle per second!) are necessary for various reasons, one of which is to minimise interference of 'dot-pattern' type which can mar received pictures due to a beat effect arising between subcarrier and line timebase frequencies. The rigid frequency relationship set up between these two gives optimum performance not only in minimising the dot pattern, but in preventing a disturbing 'crawl' effect of such pattern elements as remain. It is also necessary to have the subcarrier within the video bandwidth, and at as high a frequency as possible for minimum interference. In the UK system both the upper and lower chrominance sidebands are exploited equally, and this means that subcarrier frequency must be chosen so that both sidebands, each extending to about 1.2 MHz, can be fully accommodated within the video spectrum, as shown in Figure 3.7.

The master sync-pulse generator, then, produces an output at subcarrier frequency, and this is fed into *two* amplifiers which produce two subcarrier outputs, but with one having a 90° phase difference with respect to the other.

'Quadrature' and degrees of timing

A complete cycle of alternating current or voltage can be regarded as occupying 360 degrees – a circle! Although this statement is somewhat arbitrary it is supported by sound reasoning. Consider a generator, for example; this will, in the simplest case, rotate through 360 degrees and during that period produce a complete sine wave. If a second generator is set going but a quarter turn ahead of the first, then the sine wave produced by this will be 90 degrees out of phase with that yielded by the first one.

Since in the case of the colour television the one oscillator drives two amplifiers, the frequency of their two outputs will be absolutely identical and, moreover, the synchronism will be maintained, but always with one output 90 degrees ahead of the other owing to the deliberately contrived 90-degree phase shift introduced by the circuit elements.

Figure 3.8 highlights the situation, where the two full-line sine waves are of exactly the same frequency but 90 degrees apart in phase. The X axis is calibrated from 0 to 360 degrees corresponding to a complete cycle of the first full-line sine wave. The second one starts a little later in time; in fact, when the first has arrived at the 90-degree mark. Thus we have a direct illustration of the 90-degree difference between the two subcarriers which, remember, are derived at the transmitter from a common oscillator or generator. Hence phase can be seen to be a function of *timing* between two

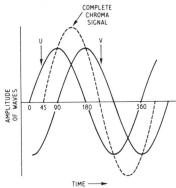

Figure 3.8. The V and U waveforms have the same frequency and amplitude but differ in phase by 90 degrees. The broken-line waveform represents the chroma signal complete which is the quadrature addition of the V and U signals

signals. The term quadrature comes from the fact that there are four 90° segments in a circle; two signals with a 90° phase relationship are said to be in quadrature.

Colour modulation

One of our subcarrier signals is amplitude-modulated with the R − Y signal, then, and its quadrature companion is amplitude-modulated with the B − Y signal. After suitable *weighting* they become V and U chroma signals of the PAL colour system. Weighting is a process of reducing the amplitudes of the R − Y and B − Y subcarriers. At a later stage they will be added to the luminance signal, and where a large subcarrier signal (corresponding to a highly-saturated colour) coincides with an extreme excursion of the luminance signal (corresponding to a very light or very dark part of the picture) the combination of the two could lead to overmodulation of the transmitter. The weighting values are: V = 0.877 (R − Y) and U = 0.493 (B − Y). After recovery in the receiver's decoder, the subcarrier signals are restored to normal proportions by simple adjustment of amplifier gain in each colour-difference signal path.

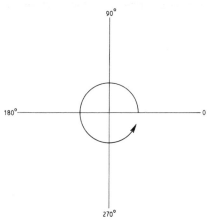

Figure 3.9. Vector 'background' diagram in four quadrants. Time or motion is regarded by convention as being anticlockwise from the 0-degree datum, as shown by the arrowheaded circle

The next move in the game is to add the V and U signals to form the chroma signal proper, and this signal is shown by the broken-line sine wave in Figure 3.8. The fundamental feature of this mode of signal addition is that by special detection at the receiving end it becomes possible to isolate the V and U signals again and thus extract the original R − Y and B − Y modulation signals. This is facilitated by the fixed 90° phase difference between the two original carriers. Signals of the same frequency and phase lose their individual identity for all time when added.

In colour TV engineering, the various colouring signals are regarded in terms of vectors. These are mathematical devices for displaying the specific features of amplitude and phase of a signal at one instant in time. The basic vector 'background' is given in Figure 3.9 where our complete timing circle is represented by the four quadrants. Motion or time is regarded as anti-clockwise, so starting from the zero-degree datum we follow the angles of phase as shown by the arrowed circle. Colour signals are generally more complex than implied by simple vectors and phasors, so this method of presentation may be regarded as a simplification, though an adequate one for our purposes here.

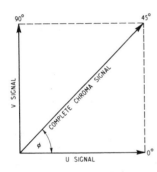

Figure 3.10. Vector representation of the V, U and complete chroma signals of Figure 3.8 (see text)

Developing the vector theme in terms of the V and U signals in Figure 3.8 we get the vector diagram of Figure 3.10. The arrowed lines here correspond to the three signals – the V and U signals and the resulting chroma signal. The amplitude of this result is obtained by completing the square (or rectangle, when V and U signals are not equal) as shown. The diagram clearly shows that the V and U signals have a 90° phase difference and that the amplitudes of these two signals are in this case equal. The angle changes (but within the same quadrant) when V and U amplitudes differ. Should the amplitudes alter together and in sympathy, then the original angle is maintained, though the vector line will now be shorter to describe lesser saturation in the transmitted colour. That the complete chroma signal is 45° relative to either the U or V signal is proved in Figure 3.8. However, it is more convenient to work with vectors than with complex waveforms and the remainder of our account will be on a vector basis.

We can easily discover the amplitude of the complete chroma signal which, in colour television at least, is often referred to as the phasor, from the expression: phasor amplitude $= \sqrt{(V^2 + U^2)}$.

Thus when the V and U signals have equal amplitude, the phasor length is 1.4 times that of V or U. Readers with some knowledge of trigonometry will also see that the angle θ of the phasor in the quadrant of Figure 3.10 is equal to $\tan^{-1}V/U$.

We next need to get clear what effect negative colour-difference signals have on the vector diagram. The quadrant of Figure 3.10 can only accommodate reds, purples and blues, and we need to describe all the colours within the triangle in Figure 2.3. To enlarge on our earlier discussion of colour-difference signals, we saw that they can move from zero in a positive or negative direction. In terms of fully-saturated colours, let's take one or two examples to illustrate this. Refer to Figure 3.11, which shows a standard colour-bar display, consisting of a white bar, followed by saturated bars of yellow, cyan, green, magenta, red and blue, then a black (zero luminance and zero colour) bar, reading from left to right. On the first bar, all three lights are on to give a full-brightness equal mix of R, G and B. There is no colour present, so the colour-difference signals are at zero as shown in the time-related luminance and colour-difference waveforms below. Moving on to the yellow bar, we know that this consists of a mixture of red and green light, so the colour-difference signals for these two will have to act to turn them up, giving rise to positive colour-difference signals, while the blue light needs to be turned off, calling for a *negative* B − Y signal. The result is the complement of blue − yellow. For the next (cyan) bar, we are looking to turn off red, and turn up blue and green (to compensate for the

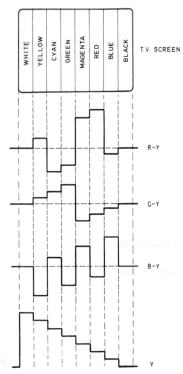

Figure 3.11. The colour-difference signals and their relationship to the TV screen display on standard colour bars

loss of *brightness* output from the now-absent red light), so here we have positive B − Y and G − Y signals, along with a negative R − Y signal. The other bars call for different combinations of colour-difference signals, but their relationship is plain from the waveforms below. Relating these to the vector diagram, let us take the case of the yellow bar. Here the B − Y signal necessarily moves in a negative direction, into the second quadrant of the vector diagram (Figure 3.12). This indicates how the chroma subcarrier reflects the polarity of the colour-difference signals in terms of its phase. In the case of yellow, the R − Y signal is still positive, so the resultant vector angle reflects the presence of a small positive value of R − Y and a large negative value of B − Y. Figure 3.12 shows the vectors for all three complementary colours, along with that for green, which (as should now be plain) calls for negative R − Y and negative B − Y signals.

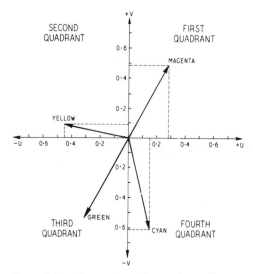

Figure 3.12. Phasors for various colours: the vector angles and lengths for green and complementary colours magenta, yellow and cyan

To summarise, however, Table 3.1 gives all the parameters so far discussed relative to the hues of the standard colour bars. All these values are based on 100 per cent saturation and amplitude.

This brings us to the concept of a 'colour clock' whose face is the vector diagram, and which has a single 'hand' able to rotate to any angle to describe the hue of the picture and

Table 3.1

Colour	Y	B − Y	R − Y	U	V	Phasor Amplitude	Angle(deg)
Yellow	0.89	− 0.89	+ 0.11	− 0.4388	+ 0.0965	0.44	167
Cyan	0.7	+ 0.3	− 0.7	+ 0.1479	− 0.6139	0.63	283
Green	0.59	− 0.59	− 0.59	− 0.2909	− 0.5174	0.59	241
Magenta	0.41	+ 0.59	+ 0.59	+ 0.2909	+ 0.5174	0.59	61
Red	0.3	− 0.3	+ 0.7	− 0.1479	− 0.6139	0.63	103
Blue	0.11	+ 0.89	− 0.11	+ 0.4388	− 0.0965	0.44	347
White	1.0	0	0	0	0	0	−
Black	0	0	0	0	0	0	−

having a 'telescopic' feature, whereby the longer the hand the greater the saturation. With the hand pointing north, as it were, fully saturated red would imply a long hand, whereas pale pink would give rise to a short one. Similarly, a bright orange colour would bring the hand to a 'long north-easterly' aspect and a green colour would direct it to the south-west. It will be remembered that the same 'telescopic pointer' effect was used in our description of the chromaticity diagram, Figure 2.3. Thus as each line of picture is scanned, so can be visualised the phasor changing in both amplitude and angle to describe the saturation and the hue of each individual pixel in the line in turn. When there is no colour, the phasor shrinks to zero and the picture is then under the sole control of the luminance part of the signal.

Integrating luminance and chrominance

We have now seen how all the colouring information is contained in a single chroma signal, by means of (a) leaving out the $G - Y$ component and (b) quadrature-modulating the remaining two colour-difference signals onto a single carefully-chosen subcarrier. The next step is to combine this with the luminance signal in a way which causes the minimum mutual interference – a daunting task, since the luminance signal appears to occupy all the available channel bandwidth! Close examination of the frequency spectrum of the luminance signal, however, reveals that most of the picture information is centred on 'energy packets' at multiples of line- and field-scanning frequencies, as intimated in Chapter 1. The spaces between these packets are relatively quiet. Let us now give some thought to the chrominance subcarrier signal. It is describing the same picture, and many 'detail' features of that picture will be common to chrominance and luminance signals. Thus the chroma sidebands will have a packet energy-distribution characteristic similar to that of the main signal, and this is the key to the interleaving process which is used. If we can *offset* the chroma subcarrier frequency to place it exactly between two of the luminance packets, as it were, the chroma

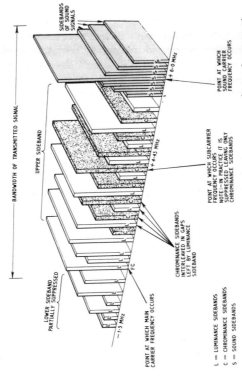

BANDWIDTH OF TRANSMITTED SIGNAL

SIDEBANDS OF SOUND SIGNALS

UPPER SIDEBAND

LOWER SIDEBAND PARTIALLY SUPPRESSED

+ 6·0 MHz

+ 4·43 MHz

~ 1·5 MHz

POINT AT WHICH MAIN CARRIER FREQUENCY OCCURS

CHROMINANCE SIDEBANDS INTERLEAVED IN GAPS LEFT BY LUMINANCE SIDEBAND

POINT AT WHICH SUBCARRIER FREQUENCY OCCURS NOTE:—IN PRACTICE IT IS SUPPRESSED LEAVING ONLY CHROMINANCE SIDEBANDS

POINT AT WHICH SOUND CARRIER FREQUENCY OCCURS

L = LUMINANCE SIDEBANDS
C = CHROMINANCE SIDEBANDS
S = SOUND SIDEBANDS

Figure 3.13. The sidebands can be looked upon as packets of energy at various frequencies. This diagram shows the interleaving of chrominance and luminance, and the situation of the sound channel in the transmitted signal

56

sidebands will fall between the similarly-spaced luminance energy packets, rather like the teeth of two interlocking combs. Figure 3.13 illustrates the principle; to achieve it the subcarrier frequency is related to the line-scanning frequency thus:

$$\text{Line frequency } (f_l) = \frac{f_{sc} - \frac{1}{2}f_f}{284 - \frac{1}{4}}$$

where f_{sc} is the subcarrier frequency (4.43361875 MHz in the UK), f_l line frequency (15.625 kHz) and f_f field-scanning frequency (50 Hz).

Suppression of subcarrier

Because the modulating information is wholly present in the sidebands of the subcarrier, it is possible to suppress the subcarrier itself at the transmitter, and by so doing we can significantly reduce the severity of the dot-pattern introduced at the receiver by the presence of the colour signal. It will be recalled that the waveform inside the modulation envelope such as that shown in Figure 3.6 results from the addition of the carrier wave and the upper and lower sidebands. As would be expected, a modulation waveform devoid of its carrier wave differs significantly from that with the carrier wave intact since the former is composed of only the upper and lower sidebands. Figure 3.14 attempts in one way to reveal the salient features.

Compared with the bottom diagram in Figure 3.6 it will be seen that the envelope has effectively 'collapsed', so that the top and bottom parts intertwine, the sine wave in heavy line representing the top of the original and that in light line representing the bottom of the original. Further, the high frequency signal inside the collapsed envelope has also changed character. The frequency, however, is just the same as the original carrier wave because it is composed of the original upper and lower sidebands, the average of which is the carrier frequency; but it can be seen that phase reversals occur each time the sine waves representing the top and bottom parts of the original envelope pass through the datum line. It is difficult to show these diagramatically; but the high-frequency signal changes phase by 180 degrees at each 'envelope crossover' point, and this has a vital significance in colour encoding, as we shall see.

The subcarrier modulation constitutes the B − Y and R − Y signals, which change continuously during a programme. However, in a colour-bar signal the signal is less 'busy' because the colour signals remain constant over each bar per line scan, changing only from bar to bar.

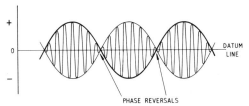

Figure 3.14. Modulation waveform when the carrier wave is suppressed. Compare this with the waveform at the bottom of Figure 3.6. An important aspect is the phase reversal of the high-frequency wave each time the modulation sine waves cross the datum line. This waveform is the addition of the upper and lower sidebands only

Figure 3.15 shows the weighted B − Y signal produced by the yellow and cyan bars at (a), the U chroma signal modulation waveform due to the bars at (b) and the reference subcarrier signal at (c). The U chroma modulation signal at (b) is with the subcarrier suppressed, and since the modulation signal is a stepped waveform going from −0.33 (corresponding to the yellow bar) to +0.1 (corresponding to the cyan bar) through the zero datum, it follows that the polarity change between the two bars will cause a 180° phase reversal of the high-frequency signal just the same as when the modulation is a sine-wave signal, shown in Figure 3.14.

The phase reversal is indicated, and more clearly shown in Figure 3.15 than in Figure 3.14. For example, it will be seen that the positive tips of the subcarrier correspond to the

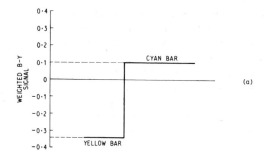

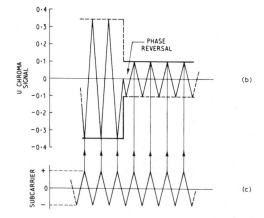

Figure 3.15. (a) PAL-weighted B–Y signal at the yellow/cyan bar transition. (b) The waveform resulting from suppressed carrier modulation. (c) Subcarrier signal in correct phase. At the receiver the reintroduced sub-carrier samples the peaks of the modulation sideband components to give an output corresponding to the amplitude and polarity of the original colour-difference signal. The polarity is given because the phase of the sideband components reverse each time the colour-difference modulation signal crosses the zero datum line (also see Figure 3.14)

59

negative tips of the U chroma signal during the yellow bar, and to the positive tips (owing to the phase reversal) during the cyan bar. Thus if we can recreate the original subcarrier at the colour receiver, and make it accurately sample the peaks of the chroma signal in this way, information is recovered on both the amplitude *and* the polarity of the colour-difference signal. This, in fact, is how the chroma signals are demodulated at the receiver. The operation of its subcarrier generator will be described in the next chapter.

Composite colour signal

When we add our carefully-contrived suppressed subcarrier signal to the luminance waveform we form the composite colour signal (CVBS signal) referred to near the beginning of this chapter. We shall have the basic television picture signal of our Figure 1.6 with the addition of the chroma signal,

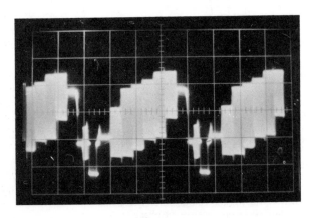

Figure 3.16 Oscillogram showing the composite signal over two lines of a colour-bar transmission

which latter will tend to ride on the luminance level, whatever that may be. Figure 3.16 indicates the form of the CVBS signal as it leaves the studio en route to the transmitting site. This is for a colour-bar signal, showing its luminance step waveform and the large superimposed subcarrier signal characteristic of fully-saturated bars. In a typical 'real' picture, saturation will not be nearly so high, and as a result the subcarrier amplitude will be correspondingly lower. The luminance signal, too, will be of a random nature, so that the equivalent composite waveform would appear fuzzy and gauze-like.

There is one other feature of Figure 3.16 which we have not yet considered, and that is the burst of what appears to be a chrominance signal on the back porch. This is in fact a reference subcarrier timing signal for the receiving decoder, and it will be considered in detail in the next chapter.

The deciphering process at the receiver

Assuming that our CVBS signal passes unscathed through the transmission system, whatever form that may take, it will arrive at the receiver's detector output (or the monitor's vision input) in the same form, and now requires sorting out to derive the original RGB signals as were present at the output from the camera. The processing of the Y signal, as in a monochrome set, consists purely of amplification as it is already in a form recognisable to a picture tube. Our concern here is to examine the overall process of colour demodulation, and see how the chrominance signal is unscrambled and recombined with luminance to recreate R, G and B signals; and finally how they are displayed on the viewing screen.

The colour subcarrier is selected from the composite signal by means of a bandpass filter, on the left of Figure 3.17. This has a response about 2.4 MHz wide, centred on 4.43 MHz to embrace the entire sideband spread of the chrominance signal. We now need to demodulate the signal, and to achieve this the subcarrier is routed to two separate switches, each capable of operating at subcarrier rate. Imagine that the switches are normally off, and that each has a storage capacitor connected to its output. In this state the stores will be empty and nothing will happen until the switches close. In fact the switches are closed for a brief instant once every subcarrier cycle. Consider the upper switch S1 which is driven by a local source of subcarrier reference. If the phasing of the reference is correct, the switch will close momentarily at the instant when the V (carrying R − Y) carrier is at its peak, and the storage capacitor C1 will acquire a charge corresponding to the instantaneous level of the V signal during the *sampling* phase. At this time the U signal (carrying B − Y information) will be passing through zero

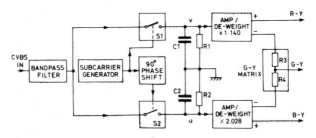

Figure 3.17. A basic decoder, showing the synchronous detector-switch system and the derivation of a G − Y signal

because of the quadrature relationship between V and U subcarriers, so that the U signal cannot affect the charge acquired by C1. If this storage capacitor is paralleled by a resistor to give an appropriate time constant, the signal appearing across the resistor will accurately follow the V signal modulated onto the subcarrier at the sending end.

The U detector works in like fashion, but here the sampling phase must match that of the U signal. This is easily achieved by introducing a matching phase shift in the reference subcarrier path to the switch, as shown in Figure 3.17. S2, then, will switch on briefly at the instant that the U subcarrier is passing through its zenith and this of course represents the time of passage of the V subcarrier through the zero datum line. The signal appearing across R2 will be a replica of the U signal at the studio.

C1 and C2 can charge to negative levels as well as positive ones, and will do so whenever the narrow and precisely-timed 'sampling window' catches the subcarrier below the zero datum line. It will be recalled that we discussed the ability of a subcarrier signal to reflect polarity changes in the colour-difference signals in terms of phase (Figure 3.15), and here we can see how the demodulators recover this information to present negative or positive signals at their outputs, dependent on the phase of the incoming subcarrier.

In practice we shall only see maximum positive outputs from the V and U demodulators on a saturated magenta picture; on the green bar, both will be at a maximum negative level, giving rise to a highly positive $G - Y$ signal which, remember, is derived by adding *inverted* $R - Y$ and $B - Y$ signals. We shall come shortly to the $G - Y$ recovery matrix. It is important that the triggering pulses for the switches, i.e. the local subcarrier signal, is precisely in phase with the broadcaster's subcarrier generator. If any timing error is present here, the sampling phases will take place at the wrong times relative to the incoming chroma signal, and this will cause incorrect or random colour-difference signal levels and polarities, making nonsense of the hues in the reproduced picture!

After detection and filtering, the V and U signals are de-weighted by passing them through amplifiers whose gain is the reciprocal of the weighting factor. thus the V signal will undergo amplification by $1/0.877 = 1.140$ to render $R - Y$, and the U signal will be amplified by $1/0.493 = 2.028$ to render $B - Y$.

$G - Y$ and RGB matrixing

Now that we have the signals back in $R - Y$ and $B - Y$ form, we can recover the missing $G - Y$ signal. It will be recalled that 100 per cent Y signal is equal to $0.3R + 0.59G + 0.11B$ which, of course, is equal to $0.3Y + 0.59Y + 0.11Y$. Now if we subtract the second expression from the first we get

$0 = 0.30(R - Y) + 0.59(G - Y) + 0.11(B - Y)$, which gives $-0.59(G - Y) = 0.30(R - Y) + 0.11(B - Y)$ and from which is obtained $-G - Y = 0.30/0.59(R - Y) + 0.11/0.59(B - Y)$. Inverting both sides of this final equation renders: $G - Y = 0.3/0.59 - (R - Y) + 0.11/0.59 - (B - Y)$. Thus to arrive at a correct $G - Y$ signal we need to add 30 fifty-ninths of an *inverted* $R - Y$ signal to 11 fifty-ninths of an *inverted* $B - Y$ signal. This is shown in Figure 3.17, where R3 and R4 select the correct proportions.

Now we are back to the three separate colour-difference signals plus a Y signal and all is plain sailing! We merely add Y separately to each of the colour-difference signals to render RGB signals ready for the display device.

Display

In our earlier example three separate display devices were used, one for each primary colour. Except in some forms of projection display this is not common, and as is well known a colour picture tube is generally used, in which all three primary colour signals are handled simultaneously. We shall cover colour picture tubes in detail in Chapter 7; suffice it here to say that the shadowmask picture tube manages to simulate the effect of three superimposed tubes, each working in one primary colour. It has a single electron-gun assembly and is set up with accelerating voltage and deflection fields in just the same way as a monochrome tube, but its three cathodes accept the RGB signals with which we started this chapter. Thus we are able to read and write in three colours to capture, transmit and display a full colour picture, with but a single link between programme source and viewer.

There is more to the mechanics of encoding and decoding the colour signal, and indeed other ways of going about it. We shall explore these in the next chapter.

The PAL system

In the last chapter, the principle of chroma encoding was explained with particular reference to the chroma signal itself, and the method of 'dovetailing' the chroma and Y signals. Reference was made to the necessity of accurately reproducing the original subcarrier reference signal at the receiver so that chroma signal phase detection can be carried out with precise timing to faithfully reproduce the hue and saturation of the colour picture. To recreate the subcarrier at the receiving end, we must transmit some reference signal to regulate the 'clock' which generates *local* subcarrier to time the sampling phases of the V and U detectors.

The colour burst

Since the subcarrier is of constant and unvarying phase, we need not send a continuous reference signal. Provided we can arrange a very stable oscillator at the receiving end (invariably a high-Q crystal oscillator) we need only send a sample for comparison at regular intervals. A place has to be found for this sample in the CVBS waveform, and the only space left which can be utilised is the $5.8\,\mu s$ back porch between the line sync pulse and the start of picture information. Onto this back porch is inserted ten cycles of subcarrier signal, with a peak-to-peak amplitude equal to the height of the sync pulse itself (300 mV in the standard-level CVBS signal). This *colour-burst* can be seen in Figure 3.16 and is the key, so far as the receiver's decoder is concerned, to the accurate regeneration of the subcarrier signal.

At the decoder, a *burst-gate* is present, triggered from the transmitted line sync pulse to open during the back porch. This isolates the burst signal and directs it into a *phase-lock-loop* (PLL) circuit which compares the frequency and phase of the local crystal oscillator with that of the bursts coming at $64\,\mu s$ intervals; see Figure 4.1. Any discrepancy gives rise to an error signal from the phase detector; applied to a reactance stage associated with the crystal oscillator it can pull the oscillator into lock with the transmitted colour burst so that its frequency and phase are identical with that at the transmitting end. Figure 4.1, then, gives the bones of a

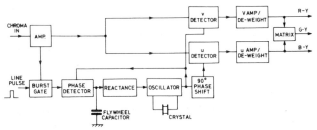

Figure 4.1. Driving the synchronous demodulators. The two transmitted components of chroma and burst are processed to render locked and correctly-proportioned R, G and B signals

complete 'basic' decoder system as an enlargement of Figure 3.17, and this basic method of encoding and decoding colour in a compatible system is that adopted in the USA, where it is known as the *NTSC* system (short for National Television Systems Committee, who recommended it for America in the early 1950s).

NTSC and PAL

The basic colour system outlined above works well, and has been in successful use in the USA for almost 40 years. Accurate decoding is facilitated by the fact that any distortions or timing errors encountered and picked up by the signal in the transmission system are imposed equally on the burst and chroma signals, so that generally their phase relationship remains constant and all is well with reproduced hues.

If we have any non-linearity in the signal's path, however, a problem can arise. We have seen that the chroma signal proper rides on the luminance signal which can be anywhere between black and peak white levels. The colour-burst, however, is always sitting on the black-level and thus passes through the system at a low level. Any source of timing error which is *level-dependent* will deal differently with the burst and chroma signals, leading to wrong-axis detection and consequential hue errors. This effect, known as *differential phase distortion*, can be quite severe, and to counter it, NTSC receivers are fitted with a hue control with which the phase of the subcarrier regenerator can be adjusted to somewhere near the correct axis, as subjectively judged on flesh-tones. This is somewhat haphazard, and certainly inconvenient, as hue adjustment is often necessary on channel-changing.

To overcome the differential phase problem, a modification of the NTSC system was suggested by Dr Walter Bruch and developed by his team in the Hanover, Germany, laboratories of the Telefunken company. This, the PAL system, has been adopted by the UK and some European countries for terrestrial broadcasting with great success. PAL counteracts any phase error which may be present on one line by introducing an equal and opposite error on the next line. The errors are cancelled electrically by an 'averaging' process in a *delay-line* matrix, to be described later, before being demodulated and used to write colours into the display with great accuracy.

Phase Alternation, Line

The scheme is achieved by the reversal in phase (effective inversion) of the V chroma signal for the duration of alternate scanning lines. What happens is that on one line of a field the

67

phase of the V chroma subcarrier is normal, then on the next line of the same field the phase reverses. These can be regarded as 'normal' and 'inverted' lines. While this is happening the phase of the U signal remains normal. *This is not phase-inverted on alternate lines.*

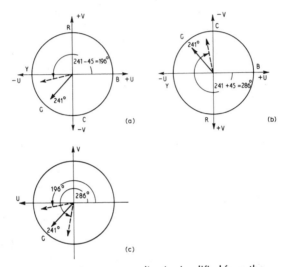

Figure 4.2. Diagrams revealing in simplified form the phase-error combating artifice of the PAL system. (a) Phasor due to green element occurring at 196° instead of 241° on a 'normal' line. (b) The same phase error on a 'reversed' line. The error is now reversed, and diagram (c) shows that the average of the two errors is 241°, corresponding to the correct phasor angle for green.

The diagrams in Figure 4.2 show how this corrects the effect of phase distortion. Diagram (a) shows a 'normal' line and phase distortion on a green element causing the phasor angle to lag 45° from the correct 241°. The full-line phasor represents the correct phase and the broken-line phasor (in all drawings) the error phase as received. The green element on this line, therefore, is displayed towards yellow.

68

Now, on the next line of the field we have the 'inverted' line which is shown in diagram (b). Notice here that the effect of the phase inversion is to invert the diagram and to reverse the direction of the phasor, so that the error now leads the correct 241° by 45°. The green element on this 'inverted' line, therefore, is displayed towards cyan.

Diagram (c) clearly shows how the average phase of the two errors (196° on the 'normal' line and 286° on the 'inverted' line) works out to the correct phasor for green, which as we have seen is 241°. In Chapter 6 we shall see how this averaging process is achieved by means of a glass delay line and an adder network.

V chroma detector switching and synchronising

Perhaps the reader has now become aware that, owing to the V chroma subcarrier phase reversals, the R − Y output from the V chroma detector would also reverse in phase unless some form of compensatory switching were introduced at that particular detector. PAL receivers therefore incorporate an electronic switch to cancel out the PAL characteristic once its usefulness has been realised; it takes the form of an inverter operating on either the V chroma signal or the subcarrier reference to the V detector − either method will achieve the desired result. The switch can be operated by pulses from the set's line timebase which will 'set' and 'reset' the switch on a line-by-line basis. More difficult are the circuit and control required to ensure that the detector switching is synchronised to the 'normal' and 'inverted' lines as transmitted, for clearly if the detector was switched to work from a 'normal' line when the input was an 'inverted' line the displayed colours would be totally wrong! Thus we need to send an identification (*ident* for short) signal to synchronise the receiver's PAL switch.

Swinging burst

This synchronising signal is conveyed by the colour bursts of the PAL signal. It will be recalled that the fundamental purpose of the bursts of the NTSC system is to frequency- and phase-lock the subcarrier regenerator at the receiver (other functions of the bursts include colour-killer switching and automatic chroma control, both dealt with in Chapter 6). The PAL bursts, however, are made to swing in phase 45° either side of the −U chroma axis in synchronism with the phase alternations of the V chroma subcarrier. This is revealed in Figure 4.3, where the burst phase on a 'normal' line is indicated at (a) and that on an 'inverted' line at (b).

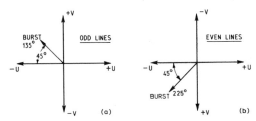

Figure 4.3. Showing how the bursts swing ± 45° relative to the − U axis on successive lines. That is, from 135° on odd lines to 225° on even lines. This means that the average phase of the bursts is 180°, coincident with the phase of the − U chroma axis. Notice how the burst swings are geared to the V chroma phase alternations

Clearly, the *average* phase of the bursts is coincident with the −U chroma axis (180°) and it is this to which the reference oscillator at the receiver locks, which is the requirement for correct colour reproduction on a PAL receiver.

The NTSC bursts do not swing in phase like this; their phase is fixed to that of the B − Y axis and is constant on all lines.

Figure 4.3 shows the 'normal' lines corresponding to odd-numbered TV lines and the 'inverted' lines to even-numbered TV lines, allowing us to refer the V chroma subcarrier and burst phases to odd and even lines. Thus from *first principles* on, say, line 7 of a field we get the conditions at (a), on line 8 the conditions at (b), on line 9 the conditions at (a) and so on.

Hanover bars

In cases of maladjustment in the receiver's decoder the PAL makeup of the chroma signal becomes visible. If alternately-transmitted colour lines are displayed in different hues due to a phase error, the fact that the TV lines are interlaced, as described in Chapter 1, means that two adjacent lines of the

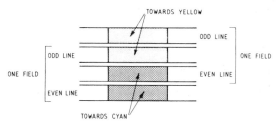

Figure 4.4. Owing to interlaced scanning, pairs of adjacent lines carry the real and opposing hue errors, and when the error is large this results in the display of Hanover bars

picture display carry a hue error, then the next two carry a complementary hue error, and so on, as shown in Figure 4.4. This gives a coarse structure to the Hanover bar pattern, and makes it more noticeable to the eye.

PAL phase identification

Because one PAL switching cycle occupies two lines, the switching frequency is at half line rate, i.e. about 7.8 kHz. This ident frequency is generated in correct phase at the receiver by a circuit which responds to the swinging bursts, and it is this which identifies 'normal' and 'inverted' V chroma signals at the receiver and maintains the PAL switch in correct synchronism. The 'PAL' switch proper is usually a pair of diodes or transistors, internal to an IC, which are alternately switched on by the action of a *bistable*, toggled by line-rate pulses and 'steered' by the ident signal.

PAL encoding system

The block diagram of Figure 4.5 illustrates the basic PAL encoding system. From the camera the RGB signals pass through a gamma correction system into a matrix, wherein the Y signal is derived according to the formula given earlier. This signal passes out of the top of the matrix block in our diagram, then through a delay line on its way to the adder.

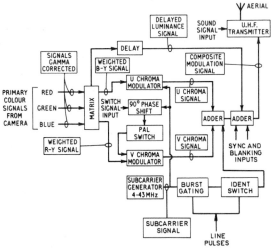

Figure 4.5. Block diagram of basic PAL transmission system. Note that U and V modulators are balanced types, which means that output from them is only present when both inputs are present

This ensures that the luminance signal emerges from the adder in time-coincidence with the chroma signals. Without this delay the luminance signal would arrive early, because the signal is delayed less as the bandwidth of a circuit is increased, and the Y signal occupies a greater bandwidth than the low-definition chroma signals. A similar delay is present in the luminance chain at the receiver for the same reasons, and must not be confused with the chroma delay line!

72

Returning now to the colour-difference signals, these emerge from the matrix and are weighted before passing into the chroma modulators, each of which is supplied with subcarrier signal; the U modulator direct from the subcarrier generator, and the V modulator via a 90° phase shift (to set up the quadrature condition) and the PAL switch (to impart the alternate line inversions). Modulator outputs are combined in an adder where the burst signal is also gated in, coming from the subcarrier generator via a gate (opened during the appropriate section of the back porch) and a switched phase-shift network to swing the burst phase through a total of 90° line by line, as per Figure 4.3.

We can see now that in essence the PAL variant differs from the NTSC signal in that alternate line phase inversions of the $R - Y$ subcarrier, and synchronised swings of the burst phase, are incorporated as a method of facilitating the cancellation of signal phase errors and hence hue error at the receiver. Let's sum up the PAL system as we have discussed it so far. The luminance signal is transmitted within the parameters of a conventional monochrome transmission. The chrominance signal is carried in the sidebands only of a subcarrier which is suppressed at the encoder; this suppressed subcarrier carries hue information by virtue of its phase and saturation information by virtue of its amplitude. To facilitate demodulation the subcarrier needs to be regenerated at the receiver, and a reference burst is transmitted once per line on the luminance back porch to synchronise the subcarrier regenerator. The V signal is phase-inverted on alternate lines to give a phase-correction-by-averaging feature, and these 'inverted' lines are identified by the swinging characteristic of the burst phase. The whole shooting-match is then sent to the transmission point, where it is reunited with the studio sound signal for transmission to the viewers' homes.

From PAL to MAC

The original NTSC colour-coding system was devised around 1950 and its PAL variant adopted about 15 years later, as was another form of colour encoding, SECAM. The latter was developed in France and is now in use in that country, the Soviet Union, and others. All these systems have the primary objective of compatibility, and all achieve it by some form of spectrum-interleaving of the Y and chrominance signals. We have seen how, in the 625/PAL mode, the colour subcarrier is carefully chosen to ensure that luminance and chrominance sidebands interleave.

What problems are introduced by this spectrum-sharing of the Y and C components of the colour TV signal? So far as the viewer is concerned, the main one is the introduction of *cross-colour*, in which luminance signals at frequencies near

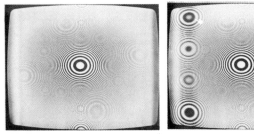

Figure 4.6. Cross-colour: the interference patterns resulting from band-sharing by luminance and chrominance signals. On the left, the signal in green as generated; on the right the display from a PAL codec system

the subcarrier are accepted by the decoder and interpreted as chroma signals. They are demodulated to produce random colour patterns superimposed on the fine-detail areas of the picture, and in subjective viewing are most noticeable on televised objects like an actor's striped shirt or a distant picket-fence, and particularly the finer definition gratings of a test card. Cross-colour manifests itself as random herring-bone patterning in red/green and blue/yellow. The effect is graphically demonstrated in the two photographs of grating-patterns in Figure 4.6. In (a) is an off-screen photograph of a green test pattern embracing virtually the entire luminance bandwidth. Photograph (b) shows the display after passing

74

this pattern through a conventional PAL *codec* (coder/decoder) system. Interference patterns are very obtrusive and the 'beat' effects, centred on 4.43 MHz, can be clearly seen.

There are several other defects inherent in the spectrum-sharing scheme such as cross-luminance (a luminance dot-pattern adjacent to chrominance changes) and the relatively poor chrominance definition we have already discussed – most noticeable on small captions, where horizontal bars of letters are coloured while vertical bars are colourless. The latter defects are not too objectionable on small screens, but would become so if and when a move is made to larger displays. Some solutions to these problems are discussed in Chapter 9.

5

Transmission and reception

At present most of our programmes come to us via the very large network of terrestrial UHF colour transmitters which has been built up since 1967 when the first UHF/colour broadcasts started on BBC2. In 1969, UHF colour broadcasting on BBC1 and ITV began, to be joined by Channel Four in 1982. In the early days, there were only a few *main* transmitters, each centrally sited in its service area, and radiating at high power. Secondary and relay stations followed, providing service in 'shadow' and poor reception areas. Today there are over 850 UHF TV transmitting sites in the UK, each radiating four channels at ERP (effective radiated power) levels ranging from 100 kW at Crystal Palace, London, to 1 W at Dychliemore in Scotland. The effective coverage of these ground stations is well over 99 per cent of the population, a remarkable achievement when it is considered that UHF waves get little further than line-of-sight!

UHF bands and channels

The terrestrial TV-transmitting spectrum is divided into two sections: Band IV containing channels 21 to 34 from 471.25 MHz to 581.25 MHz, and Band V containing channels 39 to 68, 615.25 MHz to 853.25 MHz. The sound carrier in each channel is spaced 6 MHz above the vision carrier. Each local group of channels works within a given UHF spectrum, usually 88 MHz wide. Some groups have a wider spectrum, and each group is known by a letter and colour, mainly to facilitate aerial selection, since aerials are designed to have a bandwidth appropriate to the group for which they shall be used. Thus colour TV viewers need only one UHF aerial to receive all four programmes from the *co-sited* transmitters.

Transmitter links

The sound and vision signals from the studio centre are sent over a cable or radio link to main transmitters, and to many secondary transmitters; most transmitters are in high and relatively inaccessible places. The cable link may be a co-axial or glass-fibre wideband system, often provided and maintained by British Telecom for the transmitting authorities. Where a radio link is utilised it will consist of a microwave system, with dish aerials at each end, very accurately aligned with each other. The receiving dish can often be seen about halfway up the mast at the transmitting site. Studio signals are directed through a control centre where *network* distribution is carried out; this acts as a 'switchboard' for incoming and outgoing transmissions.

Transmitter range

At UHF the distance over which the signals can be transmitted is limited to a little beyond line-of-sight, due to the curvature of the earth's surface. Thus reliable reception can only be assured within 30–40 km of a high-power transmitting aerial even when no obstructions are present. On a high receiving site 60 or more km may be achieved, though weather and propagation variations would probably affect reception.

To 'illuminate' those areas out of range of the main transmitters or shaded by local hills or mountains, *relay* transmitters are used, typically with an ERP of 1 kW or less per channel. Most work on the 'transposer' principle, in which the transmissions from the nearest main station are picked up by a suitable aerial, and each carrier is beat against a stable local oscillator to produce a third frequency which is modulated in the same way as the incoming transmission, and will form (after filtering and amplification) the 'relay output signal' for local use. This avoids the necessity to demodulate and subsequently remodulate the baseband signals, and makes for good economy and a high standard of rebroadcast signal.

Polarisation and ERP

When UHF signals are launched from a transmitting aerial, the geometry of the latter can be designed to impart either a vertical or horizontal polarisation to the electric field created in the air. By suitably polarising transmitting and receiving aerials, a great reduction in the level of interference between transmitters working on the same channel can be achieved, and this property is fully exploited in arranging the 3400 or so (850 × 4) transmitters in the UK to share the 44 UHF channels available with a minimum of interference, even in the face of tropospheric abnormalities which tend to increase the transmission range. A precision *offset* of carrier frequency is also employed to minimise the subjectivity of any co- or adjacent-channel interference which may occur.

We have referred several times to ERP in discussing transmitters, and the effective radiated power is that which is actually *radiated* by the aerial. Because transmitting aerials are designed to have a *directional* field, the suppression of radiation at some angles allows a greater concentration of energy in the wanted direction, so that a more concentrated beam is achieved, with a useful gain in transmitter range. Because all reception takes place at ground level, upward radiation into space is suppressed; many transmitters have horizontal directional properties, too, so that wasteful transmission over the sea or into a mountainside is avoided. By this means the ERP takes account of the directional power gain of the aerial system. This 'gain' characteristic is a feature of receiving aerials, too, as we shall see later.

Vision transmitter

Each station usually has a standby transmitter which comes into operation should the main one fail. In main transmitters the sound and vision signals are dealt with separately until they are in high-power RF form, when they are routed into a combining unit, from which a single co-axial cable takes the UHF signals to the common radiating aerial.

Some large transmitting stations include basic studio equipment such as slide scanners and caption generators for producing test signals or 'breakdown apology' captions. The IBA maintains a number of ROCs (Regional Operations Centre) at main transmitter sites, from which all the main and secondary transmitters in a region are monitored and controlled.

The video signals from the studio centres are received, monitored and then 'clamped'. This latter procedure is very important in television. The various components of the composite signal have to be maintained in specific proportions. The black level intervals, for example, have to hold steady at all times, as does the amplitude of the sync pulses and colour bursts. Variations in proportions and levels can occur in the signal after leaving the studio centre, and so it is arranged that the video signal is clamped electrically to a stable and fixed level before it is applied to modulate the carrier wave. Teletext signals in the field blanking interval are extracted, 'cleaned' and reinserted at this stage, then the baseband signal is ready to pass into the modulator.

As we have seen in Chapter 1, amplitude modulation is used for the vision signal as shown in Figure 1.11. The high-power AM UHF signal is launched into space by a stack of dipoles or slot elements at the top of the mast, each radiating element being spaced by half the wavelength ($\lambda/2$). The aerial's directional properties are imparted by carefully adjusting the phase of the carrier signal fed to each tier of elements. To achieve a downward tilting of the beam, for instance, a progressive phase lag is introduced into the carrier as we move downwards from the topmost tier, and in practice this is done by suitable trimming of the feeder cables to each tier. The entire radiating panel is then encased in a closed fibreglass cylinder for protection from the weather – this gives the characteristic 'candle' appearance that we see crowning the transmitting mast.

Sound transmitter

The sound signal is also brought into the transmitting station and monitored before being fed to the preamplifiers which supply the modulation input to the sound transmitter. In the UK TV system, terrestrial transmitters use FM for sound, which has several advantages over AM. The dynamic range of the audio information (e.g. the range between the softest and the loudest sounds) can be more readily retained, and because most of the interference experienced on sound reception (and indeed vision) results from amplitude variations of the modulated carrier wave, this can be eliminated by amplitude *limiting*, that is shaving off the top and bottom of the waveform at the receiver, when the sound is carried by FM. Moreover, high-quality sound reproduction is facilitated by an FM system, and compatible stereo is possible by several different techniques. The UK has settled upon the *Nicam 728* system for stereo TV sound. Here an additional carrier at +6.552 MHz carries pulse-code modulation signals containing information on 'L' and 'R' sound channels, quite separate from the established FM carrier.

Other desirable features stem from the use of FM sound in the system as a whole. For example, it makes possible at the receiver a common IF (intermediate-frequency) channel for the sound and vision signals, owing to the lack of interaction between the two types of modulation. It also facilitates the system of *intercarrier sound*, where a 6 MHz FM sound signal is produced from the beat between the sound and vision carriers, spaced as they are by exactly 6 MHz. Since the accuracy of this 6 MHz signal is ensured by the crystal-controlled references at the transmitter, it follows that the intercarrier sound scheme eliminates the effect of any tuning-frequency drift at the receiver on the sound signal – it would be far otherwise if the sound IF at 33.5 MHz were directly demodulated.

Because of its lesser bandwidth and the FM modulation system there is no need for the sound transmitter to have so great a power as its vision counterpart. In system engineering, the ideal is that in circumstances of signal attentuation, interference etc., all components of the system should have about the same threshold of failure; there is little point in maintaining a perfect sound signal when the picture is completely obliterated by 'snow' or synchronisation is lost in the presence of heavy interference. This was a guiding principle in the design of the colour-encoding system and sync/vision ratio. For sound, it has been found that the 'common threshold' is achieved when the ratio of vision to sound carrier power is about 10:1 (sound carrier 19dB below vision carrier power), and this is commonly adopted for UHF TV transmitters throughout the UK.

Vestigial sidebands

We have seen in Chapters 1 and 3 that the information transmitted by a modulated carrier wave lies in both the sidebands (see Figure 5.1a). Each sideband carries the same information (they are mirror-images of each other) and so a duplication is present, and more bandwidth than necessary is used up. It is possible to send a complete signal using one sideband only; and this is necessary in the UHF bands to conserve precious spectrum space. However, it is not possible (by filtering) to sever one sideband as cleanly as shown at (b) in Figure 5.1, so suppression usually takes the form of a 'tail' or a gradual roll-off of the unwanted portion. The choice is to have this occurring either in the wanted sideband or in the unwanted one, as shown by (c) and (d) in Figure 5.1. Neither of these alternatives is really satisfactory. In the first case low-frequency energy is lost if the roll-off occurs in the wanted sideband, while in the second case incomplete suppression takes place and the bandwidth of the signal as a whole is still fairly large. In television a compromise is sought, where the lower sideband is partially suppressed, and because there is a vestige of it remaining, the scheme is known as *vestigial sideband transmission*.

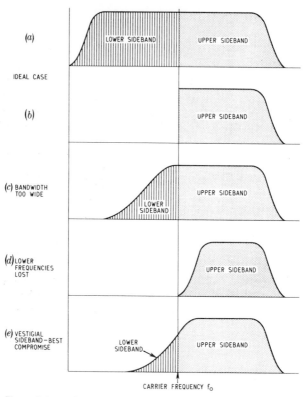

Figure 5.1. Sideband structure and vestigial sidebands. These are explained in the text

The net effect of this type of transmission is that the TV transmission uses the practical minimum of band-space, but when picked up and processed by a receiver suitably designed to accept it the assymetrical sideband energy is compensated for and equalised to give reception as good as that from a double-sideband system. The spectrum of energy for a single transmission channel is given in Figure 5.2, where the sound and vision carriers can be seen with sideband energy distribution and limits. This diagram also shows the

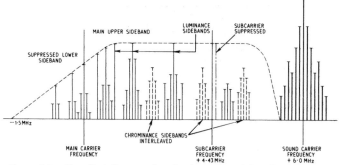

Figure 5.2. Detailed diagram showing how the sidebands of the chroma signal are interleaved with those of the luminance signal. The diagram also shows the sidebands of the sound signal and the suppressed lower sideband

interleaving effect of the luminance and chroma components of the vision signal within the upper sideband of the vision carrier.

It should be borne in mind that this is but a single 8 MHz-wide channel in the broadcast spectrum. One has to imagine the diagram of Figure 5.2 reproduced many times,

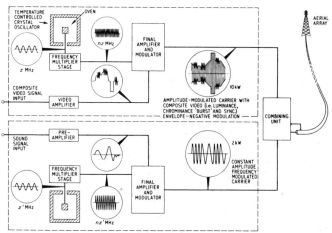

Figure 5.3. The prime elements of a terrestrial colour television transmitting station

with each laid side by side, stretching to right and left of our diagram, to get a picture of the UHF band in action. Obviously, receiver bandwith must be not only carefully shaped in its response curve, but also provision must be made to reject (by means of notch filters) out-of-band signals in adjacent channels.

To round-off our description of terrestrial transmitters, the diagram of Figure 5.3 shows in simplified form the basics of a broadcasting set-up.

UHF reception

Having launched our carefully composed RF carrier signals into the air, it's now time to examine the means of recapturing them at the receiving site and turning them back into an electrical signal for application to the TV set. Like any radio receiver the TV tuner has the job of selecting one channel from the many presented to it, and rejecting others which are out of channel. The conventional UHF aerial seen on millions of rooftops is the *Yagi* type, named after its Japanese inventor, through other types are used in rare cases.

Aerial bandwidth

Aerial elements are usually referred to in terms of wavelength, and a dipole is said to be 'tuned' when its length corresponds to one-half of the wavelength of the carrier it is to receive. Because the wave velocity reduces slightly when it is captured by the aerial element, the aerial (physical) half-wavelength is about 95 per cent of the signal half-wavelength.

At the high frequencies of the UHF channels, the active element may not be much more than 200 mm. However, to avoid attenuating the sidebands of the wanted colour signal, the aerial must have a sufficient bandwidth to pass not only the signals of one channel, but also those of the other three in the local group. As we have seen, the local group of channels may extend over 88 MHz or more. If the aerial bandwidth is insufficient or if the aerial is of an incorrect channel grouping for the area, chroma sidebands may be seriously attenuated to give severe desaturation; more often one or more channels of the group may have a poor S/N ratio

Table 5.1. Groupings and colour codes for UHF receiving aerials

Group	A	B	C/D	E	W
Channels	21–34	39–53	48–68	39–68	21–68
Colour code	Red	Yellow	Green	Brown	Black

due to uneven response of the aerial. The standard colour-codes for receiving aerial groups are given in Table 5.1.

The Yagi array

The most popular type of TV aerial to date is the Yagi design, developed in the 1920s and much used since then for its good gain and directional properties. It consists basically of a

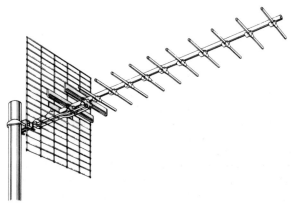

Figure 5.4. UHF Yagi array set for receiving a signal of horizontal polarisation

half-wave dipole, one or more reflectors and a large number of directors, as shown in Figure 5.4. When the array is used for reception the signal energy is 'focused' on the dipole (which of course is the active element) by the directors, and any which disperses beyond it is collected and re-radiated back to the dipole by the reflector behind it. The net result is an array with high gain (compared to a single dipole) due to its highly directional property which also confers an ability to reject signals which arrive off-beam, useful in areas like the UK where many transmitters are present, working on the

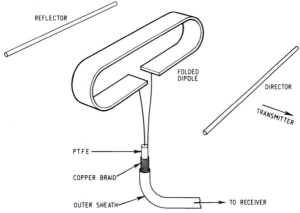

Figure 5.5. A folded dipole connected to a coaxial cable feeder. Notice the reflector behind and the director in front, relative to the direction of the transmitter

same or adjacent channels, and where buildings and natural topography can cause signal reflections.

Whether an element is to act as a reflector or director depends on its length relative to the dipole. When it is longer it behaves as a reflector and when shorter as a director. The element spacing on the boom in terms of wavelength is also of importance in determining the behaviour of an array. The gain tends to increase as more directors are added, but the law of diminishing returns eventually comes into play, and 18 elements are the normal maximum in domestic aerials. If more gain is required, two or more aerials can be used together, linked to each other, and the receiver, by a *phasing harness*. The characteristics of this, and the relative spacing of the aerial arrays, can be engineered to achieve specific requirements of gain and directivity.

The reflector and director elements are termed 'parasitic', meaning that they are not receiving dipoles as such. As more elements are added, so the centre impedance of the dipole falls, and in order to maintain a reasonable match to the feeder the dipole is folded double as shown in Figure 5.5. This artifice steps up its centre impedance by four times and thus compensates for the presence of multiple parasitic elements.

Aerial installation

While set-top and loft aerials may give passable reception in clear zones within the primary service area of a transmitter, the only reliable installation is a roof or chimney-mounted aerial with a low-loss feeder to the TV receiver. The elements must agree with the polarisation of the transmitter to be used, which generally (but not always!) means that they must be horizontal for main transmitters, vertical for relays. In situations of low signal strength, a slight upward tilt of the front 'firing' end of the array may improve results by a dB or two, though a *masthead amplifier* may be necessary to achieve an adequate S/N ratio in the picture in very weak reception areas. This consists of a small wideband transistor amplifier mounted very near the dipole to amplify the signal before further degradation occurs in the feeder cable, which now carries a DC power supply for the masthead amplifier, as well as the UHF signals on their way to the receiver.

Normally the aerial needs to be 'looking' directly at the transmitting mast, and this is generally achieved by swinging it while monitoring signal strength on a suitable hand-held meter. Sometimes, due to multipath reception and the resulting 'ghosting' effect on pictures, an orientation must be chosen for minimum spurious pickup rather than maximum signal reception.

Log-periodic aerials

An alternative to the Yagi array is the log-periodic type, which works on a different principle, although it looks superficially similar. It consists of a pair of parallel booms mounted a few centimetres apart with separate dipoles mounted at intervals along its length, their connections being transposed between adjacent elements. At various frequencies in the UHF band different dipoles come into tune due to their differing lengths, while the twin beams on which they are supported act as a transmission line, at the end of which the signal is matched into a co-axial cable. The elements adjacent to the active dipole (whichever it may be at the frequency in use) act as directors and reflector. Thus the log-periodic aerial has some directivity, but not the very narrow beamwidth of a Yagi array of comparable size; its gain is also comparatively low. The main advantage of the log-periodic type is its superior bandwidth and absence of spurious resonance effects. As such, it is useful as a measuring aerial (the BBC and IBA use them to plot field-strength contour maps), and in areas where a wide spread of grouped channels is encountered. They are also popular for use on caravans and boats where (hopefully!) reflections and signal strength problems will not be encountered, and one aerial can be expected to operate over the whole of UHF bands IV and V.

Down-lead and matching

The characteristic impedance of the dipole of an aerial system is designed to appear as close as possible to a 75 R resistor. This matches the impedance of the transmission co-axial cable which conveys the signal to the receiver's aerial socket, ensuring least loss from standing waves due to mismatching. At UHF the attenuation in the down-lead is significant and low-loss feeder is recommended to minimise this, installed to avoid sharp bends or kinks. The co-axial cable itself consists of a flexible outer sheath and an inner conductor supported through the centre of the sheath by low-loss insulating material, called the dielectric. This type of feeder has a characteristic impedance of about 75 R. The half-wave dipole in a Yagi array has its ends terminated in a weatherproof box, wherein the feeder is connected, sometimes via a small *balun* or matching transformer. In the absence of this the inner core of the co-ax cable connects to the upper dipole end while the outer sheath is bonded to the lower dipole end.

For further information on aerials, siting and transmitters in the UK, consult the Engineering Information Department of the BBC (Broadcasting House, Portland Place, London W1A 1AA, tel. 01–580 4468, ext. 2921) or the IBA (Crawley Court, Winchester, Hants S21 2QA, tel. 0962–822 444).

6

Colour decoding

Since we discussed the PAL encoding system we have strayed into RF propagation and amongst the chimney pots. Readers who feel it necessary are advised to run over Chapters 3 and 4 again before continuing here with the decoding or restoring process.

Let's first define what we are seeking to achieve. We left the chroma signal, in PAL form, riding on the luminance waveform as a modulation signal for the transmission system. This modulation signal will be rendered in baseband form at the receiver's vision detector, and the job of the decoder is to process this signal to derive R, G and B drive signals to drive the display device. Here we must digress for a moment into the terminology we are using. In the original colour TV receiver plan (at the outset of colour broadcasts) the *decoder* dealt only with chroma signals and its product was the three colour-difference signals, $R - Y$, $G - Y$ and $B - Y$. The final matrix for RGB was the picture tube itself, which summed the two signals applied to each of its guns – difference signals to the grids and the Y (luminance) signal to the cathodes, the latter sourced from an independent video amplifier just like that used in a monochrome set. Thus in some set designs the decoder could be removed from the set altogether, leaving it operating in monochrome on the Y signal. For many years now the colour-processing circuits of a TV have dealt with both Y and chroma signals together to render R, G and B signals as output; three separate primary-colour video feeds to the display device, in just the same form as they emerged from the three vidicon pickup tubes of our Figure 3.3. This is the current concept of a decoder (in the context of a PAL signal, anyway), and throughout this chapter this convention will be used.

Taking first the reference (burst) signal, we need to use this to establish within the decoder a reference timing 'clock' against which to measure the timing variations of the chroma signal itself – it will be recalled that the timing of the subcarrier is indicative of the hue being described. We need to provide an inverting switch to restore normality to the V signal on 'PAL' lines; and control it by means of the swinging alternations of the burst signal. In the chroma signal chain is provided the delay line matrix which overcomes the effects of differential phase distortion. During demodulation the subcarrier is discarded, then the baseband V and U signals are de-weighted and matrixed to derive a $G - Y$ signal. Finally, all three colour-difference signals are added to the Y component to give low-level primary-colour signals. Further amplification brings these up to the level required by the picture tube or other display device.

Chroma channel

The chroma channel is contained in the block diagram of Figure 6.1. It consists of the two chroma amplifier stages, with the ACC and colour-killer connections, the delay-line driver, the PAL delay line and the PAL matrix. The other sections, though associated with the chroma channel, will be considered later.

Now the composite signal from the vision detector, assuming that the receiver is correctly responding to a colour signal, is passed to chroma amplifier stage 1 through a high-pass filter. Since the chroma information is centred on 4.43 MHz, the high-pass filter accepts all this information while attenuating the lower-order components constituting the luminance information. Thus in essence only chroma information arrives at the input of the first chroma stage. The signal is then amplified by the second chroma stage and by the delay-line driver, these having a bandpass characteristic of about ± 1.2 MHz centred on 4.43 MHz. The skirts of the response show a fairly steep roll-off at either side, determined by the filtering in the chroma signal path to the input. Manual control of the level of the chroma signal is provided by the user colour control, while automatic control (ACC) is also provided at the first chroma stage. The second stage in the block diagram is controlled by the colour killer.

The ACC is generally applied to the stage as a potential which regulates its biasing and hence gain depending on the level of the level of the chrominance signal, as indicated by the amplitude of the colour-bursts (the only part of the

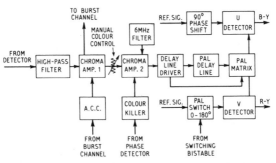

Figure 6.1. Block diagram of chroma section of colour decoder. Each section is fully explained in the text

chroma signal transmitted at a constant, known level). This reference level is picked up from the burst channel (described later) and then rectified to yield the control potential.

The second chroma stage in Figure 6.1 is deliberately designed to be non-conducting (e.g. biased off) in the presence of monochrome signals. This allows the receiver to work in black-and-white without the colouring circuits being active. If they were open during this time colour interference and noise might mar the monochrome display due to spurious colouring signals getting through the chroma channel. Now, when the colour killer detects a colour transmission by the presence of bursts, it produces a 'turn-on' bias which opens up the chroma channel, thereby letting the legitimate colouring signals through. The colour killer generally receives ripple signal (at half-line frequency; see later) from the burst-phase detector and this, or a part of it, is rectified and smoothed to produce the switch-on bias for the controlled chroma stage. A final function of the second chroma stage is that of *burst blanking*. We do not need the burst signal in the chroma chain beyond this point (it has already been diverted to the reference signal channel) and a line-rate pulse, suitably delayed, is fed to the 2nd chroma stage in time-coincidence with the burst signal to blank out the latter and prevent it from upsetting functions further downstream in the signal path, such as *clamping*.

The delay-line driver thus delivers chroma signal to both the PAL delay line and the PAL matrix, and it is this part of the chroma channel which deletes the 'phase sensitivity' of the chroma signal, and neatly separates the V and U signals for feeding to their respective demodulators.

The PAL function

We have already dealt with the PAL function from the signal point of view in Chapter 4, but since it is particularly important no apology is made for referring again to it here, but this time more specifically from the point of view of the receiver. The relevant circuit section of the PAL delay line and matrix is given in Figure 6.2. Some sets employ an arrangement just like this, while others differ in detail; nevertheless the net result is the same in all cases.

Chroma signal is driven into the delay-line input trans-ducer which converts the electrical signal into an accoustic one, whose rate of propagation through the glass block is relatively slow. Mechanical vibrations travel along a defined path within the glass block, whose edge surfaces are arranged to fold the path by multiple reflections; see Figure 6.2. The delay suffered by the accoustic wave in this long folded path is just under the period of one TV line, after which it arrives at the output transducer and is reconverted into an electrical signal by what amounts to a piezo microphone on the receiving face of the glass block. Chroma signal is also picked up from preset P1 and from this fed to the output of the line. The output transformer T2 thus receives two lots of chroma signal, one delayed in the line, and one direct. It is the job of T2 to add and subtract (i.e. matrix) direct and delayed lines of chroma signal. These processes, it will be recalled, effectively remove the 'phase modulation' from the chroma signal.

It works like this. We get addition and subtraction because the transformer T2 is bifilar wound and centre-tapped. The effect is that the signal from the delay line across winding A is

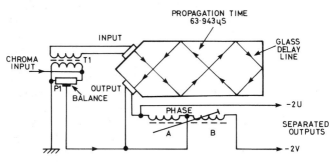

Figure 6.2. The delay line and its matrix. The folded accoustic path within the delay line is arranged to take 63.943 μs to traverse

exactly 180° degrees out of phase with the signal also across that winding from preset P1, while the signal from the delay line induced into winding B is exactly in phase with the signal also across that winding from preset P1. The antiphase signals across winding A thus cancels out, assuming equal strengths (one is subtracted from the other), while the in-phase signals across winding B add together.

Let us first consider the action on the U components of the direct and delayed signals. The subcarrier upon which the U signal is based is not phase-reversed at any time in transmission, so it has the same base phase on every line. Since the delay is engineered to be 63.943 μs it holds 283½ cycles of reference subcarrier, and the emerging signal is half a cycle out of phase with the direct one – this represents a phase relationship of 180° so that when direct and delayed U signals are *added* they will cancel out thus: $-U + U = 0$. From the subtracting function, we get: $-U - U = -2U$, assuming equal strength direct and delayed signals of unity value. The left-hand side of winding A on T2 in which subtraction occurs is thus endowed with a 'pure' U signal (albeit negative) and no V signal at all.

The V components of the direct and delayed signals are similarly processed, but the net result is different because (see Chapter 4) the transmitted subcarrier upon which the V chroma is based is reversed in phase during alternate lines, which means that the V components themselves are similarly processed. The adding part of the matrix thus 'sees' V signal of, say, 'positive phase' via the delay line and V signal of 'positive phase' via the direct route. If this is not clear, remember that 283½ cycles in the delay line gives a phase inversion between input and output. With the transmitted signal having also undergone a phase inversion since the last line (PAL feature), the two inversions cancel out *on every line* so that the adder (T2 section B) will produce $V + V = 2V$ on every line, while the subtractor, section A, will always produce $V - V = 0$. So it is that the right-hand side of T1 section B will produce pure V signal with no crosstalk from the U channel.

To summarise, therefore, the U detector receives only U chroma signal because the alternate-line phase reversals of the V chroma signal result in cancellation of the V chroma components in the subtractor section, while the V detector receives only V chroma signal because the in-phase V components reinforce in the adding section while the antiphase U components are added and thus cancelled. In other words, the composite chroma signal is very neatly separated into its two V and U constituents, these now carrying information in terms of amplitude *and not phase*. This means that phase distortion (which is tantamount to 'timing errors') in the system can no longer introduce hue errors on the display. If differential phase errors are fairly large there is a mild by-product in the form of reduced saturation, but this is more subjectively tolerable than a change of hue!

We still need to introduce the reference signals to the V and U chroma detectors (their operation was described in Chapter 3 and illustrated in Figure 3.17), in phase quadrature of course, to recover the quadrature-modulated baseband V and U signals. We must also neutralise the effect of the $\pm$ V chroma signal at some stage. In Figure 6.2 P1 adjusts the matrix balance so that the direct and delayed signals have equal amplitude. Imbalance here can give rise to Hanover bar effects as described in Chapter 4.

We can now return to Figure 6.1 again and see that the reference signal to the U detector undergoes a 90° phase shift which provides the quadrature condition just noted. It may be recalled that in our simple illustration of the delay line matrix system we showed the U output as being in *negative* form. ($- 2U$ at subtractor output) and to correct this we would need to either invert the U signal before detection or *introduce a further 180° phase shift in the reference signal to the U demodulator.* Either will achieve the desired effect of rendering, at the U detector output, a correct-polarity signal. In the case of the V-detector we have to neutralise the phase inversions of the transmitted signal, and as described for U above, this can be achieved by either inverting the V signal itself, or the subcarrier feed to its detector, but this time on a line-by-line rather than a permanent basis. The switch thus operates effectively at line timebase frequency, but remember that this is the 'toggling' rate; Figure 6.1 shows the switch in the reference signal path.

The PAL switch commonly consists of a pair of diodes or transistors within an IC, switched alternately to opposite phases of the local reference signal, the actual switching being governed by a bistable generator (this is of the multivibrator family but is stable in both switching modes and requires a trigger pulse to switch it from one state to the other) triggered by pulses from the line timebase or sync separator. It thus switches state line by line, each state having a frequency equal to half the TV-line repetition frequency.

Provided the switching count corresponds to the plus and minus phases of the V chroma signal, therefore, the V detector will work as though it is receiving a V chroma signal of constant base phase; but if the switching count is in error the V detector will fail to work correctly and the displayed hues will be in error. To ensure that the switching is correctly synchronised to the transmitted V chroma phase inversions, line-identification pulses are also fed to the bistable from the phase detector responding to the swinging bursts, but this story is yet to come.

Burst and reference signal channels

Another section of the PAL decoder is given by the block diagram of Figure 6.3. Here the burst gate and amplifier receives composite chroma signal from the chroma channel. The gate part is a transistor or diode which is switched at line frequency such that it conducts only during the periods of the bursts; during the rest of the television line it is non-conducting, thus performing the opposite function to the burst-blanking switch in the chroma amplifier described earlier. This means, then, that only the bursts of the chroma signal are amplified and that the output from this 'block' consists only of a series of bursts with no picture content.

Note that the bursts are also fed to the ACC circuit in Figure 6.1 which, after rectification and smoothing, provide the control bias for the chroma amplifier, already considered. There are many possible causes of chroma-signal level variation, but the ones we are most concerned with are those where the chroma signal is affected to a different degree than the luminance component. The resulting imbalance would upset saturation if it went uncorrected by some form of ACC; the most common causes of this sort of imbalance are RF mistuning and aerial or propagation defects which unduly emphasise one section of the RF channel bandwidth.

Returning to Figure 6.3 and the bursts, however, they are primarily fed to a phase detector which compares their average phase (remembering that their 'swinging' character-istic takes them alternately ±45° relative to the − U axis; see Figure 4.3) with the phase of the locally-generated reference signal. This reference signal for such phase comparison is fed via a 'sample loop'.

The reference generator is crystal-controlled but mild phase variation of the reference signal is made possible by a varicap diode effectively in shunt with the crystal. The varicap diode is widely used in applications where a DC-controlled reactance effect is required. It consists of a p-n semiconduc-tor junction operated in reverse-bias which, under this condition, exhibits a capacitance effect; the effective value

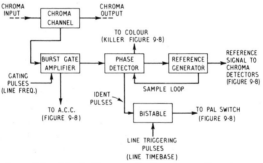

Figure 6.3. Block diagram of reference and burst channels

decreasing as the reverse bias is increased. This happens because the *depletion layer* between the heavily doped p and n regions acts as a dielectric and its width is a function of applied reverse bias. Thus we can vary the crystal's shunt capacitance, and hence its phasing, by simply applying a DC voltage to the varicap device.

Now, for correct chroma detection it is essential that the phase of the reference signal applied to the detectors corresponds exactly with the phase of the subcarrier which was suppressed at the transmitter. The 90° phase shift required between the two detectors is provided by the phase shifter in Figure 6.1, which has already been mentioned.

The phase detector

It is the job of the phase detector to provide a control potential for the varicap diode such that the phase of the reference signal is always held constant. This is easily possible because the average phase of the bursts corresponds exactly to the average phase of the subcarrier at the transmitter. The phase detector, in fact, produces a DC output of a positive or negative value corresponding to the phase error between the bursts and the locally-generated reference signal. This is called the *error* voltage, and it is superimposed upon the standing reverse bias applied to the varicap diode so that the phase of the reference signal is continuously held at the correct value, with any tendency to drift being immediately corrected. Thus is set up a phase-locked loop (PLL) in which the output signal is locked or *slaved* to the incoming reference, in this case the colour bursts. Modern electronic equipment makes much use of PLLs.

The correctly-phased reference signal is then fed to the V and U chroma detectors via the sections shown in Figure 6.1. Some designs will have a DC amplifier between the phase detector and the varicap diode, while there is always a 'buffer' stage (probably an *emitter-follower*) between the reference generator and the chroma detector feeds.

Ident pulses

We have seen (Chapter 4) that the bursts are swinging in phase line by line such that on one line the phase relative to the − U chroma axis is + 45° and on the next line − 45° and so on. These swings are geared to the phase reversals of the V chroma signal to provide a facility for the identification of the + V and − V lines (odd and even lines, Chapter 4) of the chroma signal, thereby allowing the PAL switch to be correctly synchronised.

So far as the phase detector is concerned the swings of phase are processed as phase modulation, which means that the phase detector's output contains, superimposed on the error voltage, a ripple at half-line frequency (about 7.8 kHz) – *half* line frequency because the phase is the same on every *other* line.

This ripple can be amplified and squared-off to produce ident (short for identification) pulses which are fed to the PAL switch bistable. It is arranged that when the line-frequency triggered bistable is correctly synchronised (and there is only a 50/50 chance at the outset!) the ident pulses will have no effect, but if the bistable is out of agreement with the transmitted V chroma the ident pulse will inhibit its operation for one count, so that it 'misses a beat', as it were. Thereafter the bistable phase will be correct and the ident signal will not be required until a new channel is selected or the set switched off and on again.

Video section

Having arrived at colour-difference signals once more (at the output terminals of the V and U detectors), we are now ready to process them towards the production of RGB signals. The first step is amplification in the R − Y and B − Y preamplifiers whose gains are adjusted to 'de-weight' the signals as described in Chapter 3, and the relative gains are 1.140 for R − Y and 2.028 for B − Y. Next comes the matrixing section for G − Y (see Figure 6.4) and a *clamping* process. This is

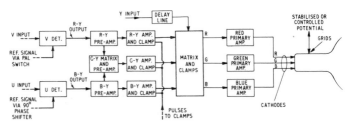

Figure 6.4. The 'back-end' of the decoder showing the stages of signal processing between the chroma detectors and the picture tube

similar to that described earlier for TV signals on arrival at the transmitter, and ensures that all three signals, $R - Y$, $G - Y$ and $B - Y$ are sitting at the same reference level so far as their black level is concerned. This is essential for the next stage, which is the coming-together of the luminance and colour difference signals to form primary-colour signals.

Let us see how this final process is done. Again it is a simple algebraic function in which the Y signal is *added* to each colour-difference signal, having itself undergone a similar clamping process. So $(R - Y) + Y = R$, $(G - Y) + Y = G$ and $(B - Y) + Y = B$. When the difference and luminance signals are added, then, we are back to primary-colour signals. For the standard colour bars of our Figure 3.10, the primary-colour waveforms are shown in Figure 6.5. In

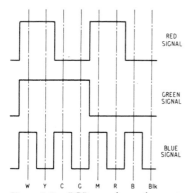

RED
SIGNAL

GREEN
SIGNAL

BLUE
SIGNAL

W   Y   C   G   M   R   B   Blk

Figure 6.5. RGB waveforms for a standard colour-bar signal. These result from matrixing Y and individual colour-difference signals

Chapter 4 we saw (Figure 4.5) that to achieve time-coincidence between narrowband chroma signal and wide-band luminance signal at the transmitter a delay line was inserted in the Y channel. At the receiver's decoder a similar delay is required to prevent the chrominance being printed slightly later (and hence to the right, as TV picture scanning is from left to right) than the corresponding luminance picture.

The delay required is of the order of 300–800 ns depending on circuit design, and the wideband delay line can be seen in Figure 6.4. It is made of series L and parallel C components, rather like a π filter; its cut-off frequency, however, is way above the highest luminance frequency in the video signal.

At the point of emergence from the matrix the RGB signals are at low level, perhaps 4–5 V black to 'white' and we shall wish to drive a colour picture tube with them. To provide the 90 V or so drive required by the picture-tube guns, each primary colour signal is passed through a wideband amplifier, very often a complementary type using two or more transistors in each channel. The characteristics of the three primary-colour output amplifiers must be closely matched in terms of gain, black level and bandwidth to ensure that the primary colours agree or 'track' at all brightness levels and frequencies. This will achieve correct relative and combined colour fidelity at all brightness, hue and saturation levels.

An IC decoder

The entire PAL decoder function is contained within a single IC or 'chip' in current receivers. These use a mixture of analogue and digital functions and require a minimum of external components. Typical of such a decoder chip is the 24-pin plastic-packaged Mullard TDA3562A illustrated in Figure 6.6. In broad outline, it follows the block diagram and description already given, but several detail differences and new features will be noticed, and these will be described in turn.

Luminance chain
The delayed luminance input (pin 8) is AC-coupled to permit clamping within the IC. An amplifier and line-rate black-level clamp is first encountered, then, followed by insertion of a reference black level generated within the chip. The insertion period (3L) occurs once per TV *field* for reasons

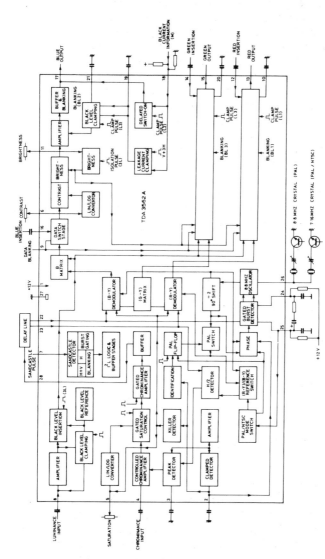

Figure 6.6. Block diagram of the TDA3562A chip (Mullard)

which will become clear later. Luminance signal then passes direct to the R, G and B matrices, and there we will leave it for the moment.

Chroma path
Chrominance signals enter the chip on pin 4, sourced from the same point (vision detector) as the luminance signal described above; in place of the latter's wideband delay line, they will have come via a high-pass filter, and so will consist of the baseband signal spectrum between about 3.3 and 5.5 MHz to embrace the chroma signal bandwidth. The first step is the ACC function, carried out in the controlled chroma amplifier – controlled, that is, by the potential derived from a peak detector and stored in the capacitor on pin 3. Next comes a gated saturation control stage under the influence of a DC control voltage from the user saturation control. The 'gated' function is puzzling here. In fact the gating prevents the saturation control from having any effect on the burst signal, which thus passes, at a constant level, though the chroma chain, delay line and all. The gated chrominance amplifier increases the chroma/burst amplitude ratio by conferring about 12 dB extra gain during the 'picture chrominance' period of 52 μs. An impedance buffer follows, whence the signal leaves the chip on pin 28 en route for the delay line.

Emerging from the delay line, the separated U and V signals now enter the chip on pins 22 and 23 respectively and are passed to the demodulators. The burst is also present, remember, and this is extracted from the delay line outputs for use in the gated burst detector at the bottom of the diagram.

Reference chain
The reference oscillator in this design runs at exactly twice subcarrier rate – 4.433619 $\times 2 = 8.867238$ MHz and this is the nominal frequency of the crystal connected to pin 26. Let us explore the reason for this double-frequency technique. We know that the two chrominance detectors require switching pulses with a quadrature relationship, and ordinarily this

would require a 90° phase-shift network of some kind in the reference path to the U detector. If we have a reference signal at $2f_{sc}$ we can halve its frequency by use of a bistable counter rather like the ident flip-flop described earlier. If we arrange two such flip-flops, one counting on negative flanks, the other counting on positive flanks, we shall have two 4.43 MHz squarewaves as outputs from them; essentially these outputs will have a 90° (quadrature) phase relationship as required for the synchronous detectors. Figure 6.7 illustrates the technique.

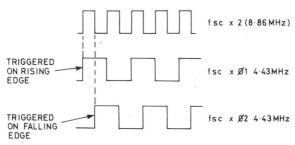

Figure 6.7. Derivation of quadrature drive signals for the chroma demodulators. A crystal oscillator generates a squarewave at 8.86 MHz ($2f_{sc}$) and a pair of $\div 2$ bistables, respectively triggered from rising and falling edges, produce two f_{sc} drive signals in quadrature

Back now to the bottom of Figure 6.6. The $\div 2$ 90° shift block contains the bistables just described and one output operates the $B-Y$ demodulator. The second output is inverted on a line-by-line basis in the PAL switch on its way to the $R-Y$ demodulator whose baseband output is thus restored to $+(R-Y)$ on both 'PAL' and 'normal' lines. The gated burst detector operates only during the back porch, comparing the $R-Y$ axis reference signal at 4.43 MHz with incoming burst to render an error voltage at pins 24 and 25, between which are connected a time-constant RC circuit to act as 'flywheel', smoothing out the ripple due to the swinging burst. The smoothed error voltage is applied to an internal varicap diode to 'pull' the crystal oscillator into lock as described earlier. Thus is the PLL set up, and the demodulator feeds are now slaved to incoming burst and held rigidly on the correct phase vector axes for correctly-timed detection and sampling.

Ident circuit
The operation of the ident detector is a little different in this chip from the established practice. Here we have an H/2 detector to compare the subcarrier output of the PAL switch with the R − Y axis component of the burst signal. Let us look a little deeper into this. We have seen that in a properly set-up delay-line matrix the V and U outputs contain only *amplitude* information, with phase variations already taken care of (page 95). Considering then the effect of the burst signal on the V (R − Y) signal, we can see (Figure 4.3) that on a 'normal' line a positive voltage will appear in the V channel during the burst period, then during an inverted or 'PAL' line a negative voltage will be present during the burst period.

When the H/2 detector discovers a discrepancy between R − Y signal phase and R − Y reference phase, the PAL flip-flop gets a 'reset' pulse from the H/2 detector via the ident stage; this corrects the switching sense and ensures agreement between the PAL switch and the transmitter.

ACC and colour killer
Once the PAL switching phase is correct, the output voltage of the H/2 detector is directly related to burst amplitude and so becomes the feedback signal in the ACC loop, finding its way back to the controlled chroma amplifier in the form of a DC control voltage. In the absence of a burst signal killer action is evoked with the chip, taking place at *two* points in the signal path to ensure that spurious chroma suppression is adequate. The saturation control input is taken low and the synchronous detectors are simultaneously cut off.

Control pulses
For economy of lead-out pins, all control pulses enter the chip on its pin 7, and three basic pulse streams are required. A field pulse is present at 20 ms intervals along with a 12 μs

line blanking pulse at 64 μs intervals. During the latter, and coinciding with the back porch, a further pulse, about 4 μs wide, is added at a higher level (hence the *sandcastle* shape of the composite pulse). These are separated within the chip by 'level slicers', then timed and processed for many different functions as the chip diagram makes clear.

NTSC capability
In view of the international market intended for such devices, the TDA3562A has a dual-standard feature in that it can also work with NTSC signals. By a clever arrangement of mode-switching and 'dual-personality' of some lead-out pins, the PAL switch can be disabled and a tint control brought into operation by DC mode switching at pins 24 and 25 of the chip. If the NTSC subcarrier frequency differs from 4.43 MHz (as it does in the USA, but not in certain 'hybrid' signal formats used in the UK in connection with special VTR playback modes, etc.) the transistor switches at the bottom of the diagram can be changed over to bring in a suitable reference crystal – the one shown is 7.16 MHz, twice the 3.58 MHz USA standard subcarrier frequency.

In combination with another chip, the device can also handle SECAM signals!

RGB video stages

It is in the post-demodulator stages that modern IC designs differ most from the traditional model of signal-processing practice. In Figure 6.6 only the blue channel is shown in its entirety – the R and G stages are identical. We left the Y signal at the RGB matrices, and it's time now to take up the story from there. The output from the matrix is of course in primary-colour form and it first encounters the *data-switch*. For each primary-colour channel this forms a double-pole switch able to select either broadcast or locally-generated signals under the influence of the data blanking input on pin 9. When pin 9 is raised to logic level 1 the switch operates and locally-generated video data (digital signals, typically from a teletext or Prestel decoder, or a computer) is passed from the 'insertion' pins (12 14 and 16 for RGB respectively) to the primary-colour outputs. The data blanking input can operate at video rate if necessary, enabling superimposition of text or graphics on analogue-derived pictures; a good example is the 'mix' mode in text-equipped receivers. As with luminance and chrominance signal inputs to this IC, the data is AC-coupled in, then clamped within the chip.

Next comes the contrast control stage, and here again we have a DC control input feeding a voltage-controlled amplifier so that signals are not routed to and from the control panel, and facilitating remote cordless command of the set's functions. Because the voltage-controlled atten- uator with the chip has an essentially anti-log control input/signal output relationship, and in view of the fact that linear control functions are easier to arrange from panel controls or remote-command decoders the chip contains at pin 6 a lin/log converter – a similar device is present at pin 5, saturation control input. Because contrast is independently but simultaneously controlled in all three primary colour channels, matching between them must be very good; it can be seen that the same applies to the next (brightness control) stage, and that inserted RGB signals in the data blanking mode also come under the influence of contrast and brightness controls.

The brightness control works on a clamping basis, restoring the three black levels to a point set by the DC control voltage on pin 11. This is carried out during the '3L' period of black level insertion at the luminance amplifier, and described earlier.

Automatic grey-scale tracking

As we shall see in the next chapter, a colour picture tube contains three separate electron guns, one for each primary colour. For correct picture reproduction in black-and-white or colour it is essential that these three guns have exactly the same emission characteristics. If the cut-off points (beam- extinguish voltage) vary between them, for instance, the same signal applied to all three will give different results on each, so that we may find red-tinted low lights because the red gun has a higher (cathode) cut-off voltage than its companions. Alternatively, perhaps, the green gun will cut off at a lower cathode voltage than red or blue, leaving a magenta tint in place of dark grey in the display. These differences, initially due to manufacturing tolerances, can be easily compensated for by pre-set adjustment of black-level voltage for each individual gun. As the tube ages, however, differential drift of the gun characteristics takes place; this necessitates occasional resetting of the internal black-level controls during the life of the tube to maintain tint-free reproduction. For some years technology has been available to automatically adjust the black-level points for each individual gun, but in discrete-component form it has been too expensive to incorporate in most domestic receivers. The TDA3562A has built-in auto black-level correction and we shall now examine the operation of this.

During TV lines 22, 23 and 24 and their companions 334, 335 and 336 on the twin-interlaced field, the scanning spot is at the top of the screen and out of sight, being just about to 'light up' to start describing the first active TV lines of the field. It is at this time that the auto-grey-scale compensation takes place. For the duration of these three lines the 3L insertion-pulse blanks all video information in the RGB channels; see the time-related waveforms of Figure 6.8. A 1-line duration clamp pulse is now inserted into each of the colour channels in turn; red, blue then green corresponding to L1, L2 and L3 in Figures 6.8 and 6.6.

This clamp pulse drives each gun of the picture tube just to the cut-off point and the beam current in the tube (nominally zero at black level, in practice a few microamps for measurement purposes) is measured and passed into the chip on its pin 18. Thus pin 18 (row 8 in Figure 6.8) receives three pulses in quick succession, sequentially giving information on the cut-off points of the R, B and G guns of the tube in turn. These reference levels are gated to the three storage capacitors connected to pins 10, 20 and 21 respectively, and the charge stored in each of them is proportional to the

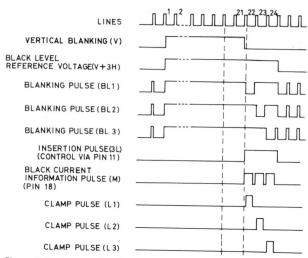

Figure 6.8. Time-related auto-grey-scale-shift and blanking waveforms for the TDA3562 IC. This diagram is centred on the field blanking interval

cut-off point of the corresponding tube gun, which will vary with temperature and aging of the picture tube.

The charge on the storage capacitor is made to modify the amplitude of that gun's clamp pulse so that a separate feedback loop is maintained for each gun; with all three guns thus controlled, automatic grey-scale correction is achieved, and the need for a black-level adjustment preset in each channel is eliminated.

RGB output amplifiers

If the display device is a picture tube, either a shadow-mask device or three separate display tubes in a projection set-up, we need three matched-voltage amplifiers to raise the RGB signals to a level sufficient to drive the modulating electrodes, usually the cathodes. The bandwidth of these needs (if data and other video sources are to be handled) to

110

be in excess of the 5 MHz or so of normal Y broadcast signals because rise and fall time (the time taken for a video signal to raise itself from black level to white, and vice-versa) is inversely proportional to bandwidth, as we saw in connection with Y/C registration and Y delay lines. Where data is being displayed the sharpness of the vertical edges of the characters depends on RGB bandwidth, the greater the

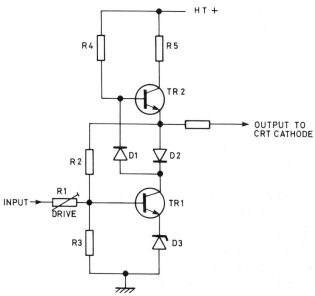

Figure 6.9. A wideband video amplifier for R, G or B channels. Circuit operation is explained in the text

better! As previously mentioned, the three amplifiers need to be closely matched in all respects as they separately handle components which are reblended in the display to form a composite and very critical image.

A typical video drive amplifier is shown in Figure 6.9. Here the video amplifier is Tr1 working in common-emitter mode with a gain of about 25. The video input signals to this stage are positive-going for 'white' and the polarity inversion in the stage renders negative-white output signals suitable for cathode drive of the picture tube. The difficulty in arranging the output amplifier is that it has to drive a wideband signal into the considerable capacitance of the tube cathode and its associated wiring; this calls for low output impedance. On transitions from black to white Tr1 is being driven on and its

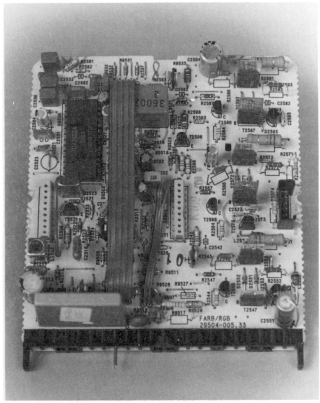

Figure 6.10. A Grundig decoder module. This one takes composite video at low level and processes it to tube-drive stage. The TDA3562 chip is visible at top left with the 8.86 MHz crystal below. Delay line for luminance is top centre, and for chrominance bottom left. On the right is a column of transistors; these form the RGB output stages

low impedance quickly discharges the stray capacitance to give a fast fall-time to the output waveform and a sharp leading vertical edge to the televised object, be it a data character or picture highlight.

When the transistor turns off, however (trailing vertical edge of picture object), the stray capacitance can only charge relatively slowly via the amplifier's load resistor R4 and because this RC time constant is long a trailing smear effect will be manifest on the picture. This is where Tr2 comes into operation, acting as a low-impedance emitter-follower with its base bias coming from R4. The shunt capacitance quickly charges via Tr2 and low-value R5, then, and crisp pictures are ensured by the low output impedance, and hence great bandwidth, of this video output stage. Two further identical amplifiers are needed in a practical TV, of course, and they share the emitter-potential reference diode D3 and work from a common HT supply of about 200 V.

One other aspect of the RGB stages must be mentioned, and this is the necessity for drive adjustment. In Figure 6.9 R1 is made variable and acts as a 'contrast' control for the primary colour in question. The presence of similar controls in the two other channels enables us to adjust the *drive* for reasons which will become plain in the next chapter. Where no auto-grey-scale correction is provided by the decoder chip, a further preset is required, this time to set the black-level for each gun. It effectively alters the standing voltage at the picture tube cathode to facilitate manual matching of the cut-off points of the three guns.

Decoder summary

In this chapter we have seen the 'chroma unscrambling' procedure in some detail, both from a theoretical point of view and in terms of modern hardware. A photograph of a decoder module using the TDA 3562A chip is shown in Figure 6.10.

7

Colour display devices

A brief account of the operation of a monochrome picture tube was given in Chapter 1, and in this chapter we shall enlarge on that description to take in colour tubes and finally other types of colour display. A great deal of research and development has gone into the technology of colour picture tubes over the years, and direct-view types are now available with screen diagonals up to 89 cm. Improvements have been made in screen materials, gun technology and energy demand. The colour tube uses all the techniques of its simpler counterpart, and by extension and refinement of these is able to present a full colour picture on a single 'integrated' screen – no mean achievement! All direct-viewing colour tubes work on the shadowmask principle originally brought to fruition by Dr A. N. Goldsmith and his research team in the American laboratories of RCA Ltd in 1950. The original device was rather clumsy and cumbersome by today's standards, but it embodied all the fundamental features of the current generation of colour tubes, and was amenable to mass production with all the cost advantages that could bring. The practical realisation of a relatively compact and decidedly cheap domestic colour TV receiver was (and is) dependent on the direct-viewing shadowmask tube concept.

The three-gun picture tube

Colour television picture tubes have to carry three different screen phosphors in order to produce three displays in perfect registration, one each in red, green and blue. If the face of such a tube (when operating) is examined under a magnifying glass the phosphor-dot formation might be seen, as shown in Figure 7.1. The phosphor make-up of the dots, of course, looks the same for each group of colour, and it is only when the phosphors are stimulated by the electron beams that the different colours are produced.

There thus exists a multiplicity of 'groups' of the three colour phosphors, and each group, detailed in Figure 7.1, is called a triad (*group of three*). In a typical tube there are almost 500 000 triads, together forming an interwoven pattern over the entire screen area.

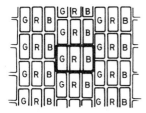

Figure 7.1. The formation of the red-, green- and blue-glowing phosphor dots on the screen of the in-line shadowmask picture tube. The three-dot formation is called a triad

The three guns are arranged in the tube neck so that the electron beam from one energises only the red phosphors, that from the second only the green phosphors and that from the third only the blue phosphors. For this reason the guns and beams are often referred to as red, green and blue. They are not coloured, of course! When a group of triads is caused to glow at proportioned colour intensities, the eye is deceived into perceiving that the three primary colours occur at the same point and so it discerns a spot or pixel of a colour corresponding to the mix of the three lights. When the mix is correctly proportioned for white light, the spot or pixel appears without colour as a 'shade' between white and black – i.e. a shade of grey.

Remembering our description in Chapter 2 of additive mixing by lights, it is easy to see how the spot or pixel can have any of the three primary colours red, green or blue; a complementary colour, yellow, cyan or magenta; or indeed any colour of intermediate value that can be produced by an admixture of the three primaries available, and falling within the 'television triangle' of the chromaticity diagram, Figure 2.3. This embraces a wide range of colours which make the colour TV display virtually subjectively equal to a good-quality colour photographic print.

The diagrams of Figure 7.2 give a basic impression of the working of the shadowmask three-gun picture tube. It is, in fact, three 'tubes' in one glass envelope and sharing the same

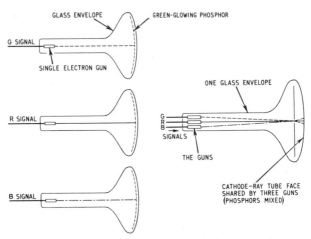

Figure 7.2. These diagrams show how the three-gun tube can be regarded as 'three-tubes-in-one'

screen. The three electron-gun assemblies work quite independently to produce the three separate electron beams, each of a different strength according to the primary-colour signals by which they are being controlled.

Shadowmask

In addition to the components of the tube already mentioned, there is what is known as a shadowmask, from which the tube gets its name. In appearance this is something like a metal gauze pierced by a large number of holes. In fact there is one hole, or slot, for each triad and the purpose of the mask is to ensure that the phosphor dots of the different colours are hit only by the electron beams from the guns of corresponding colours. This will made clearer shortly!

The shadowmask is located about 12 mm behind the phosphor screen, and the three beams are made to converge on a single hole and cross over within it to diverge slightly beyond so that the red, green and blue beams strike their corresponding fluorescent colour phosphor dots.

Each dot is then activated by the electroluminescent process described in Chapter 1 to yield an amount of coloured light proportional to the intensity of the electron beam striking it. As already mentioned, the closeness of the dots and the integrating action of the human eye merges the three individual colours into a single spot whose colour and intensity are closely controlled by the video signal. A broad impression of the total TV illusion is given in Figure 7.3 which illustrates the simultaneous analysis, passage, reassembly and perception of a single pixel, of which many millions are sent every second!

The shadowmask is made of a thin sheet of metal identical in size, shape and contour to the glass faceplate of the tube. It is manufactured by a photochemical etching process similar to that used for the production of lithographic half-tone plates for printing. The positional and etch-timing accuracy has to be several orders higher for the shadowmask, however!

Arranging the phosphor dots

To produce the phosphor pattern on the inner surface of the glass faceplate a very ingenious method is used. The green glowing phosphor material is first sprayed onto the screen so that it covers the whole area, and then the shadowmask is mounted accurately in the faceplate. To 'fix' the green phosphor material on the screen in the dot-pattern required, ultra-violet light is arranged to pass through the shadowmask holes at an angle corresponding to that of the electron beam from the green gun when the tube is correctly active. The

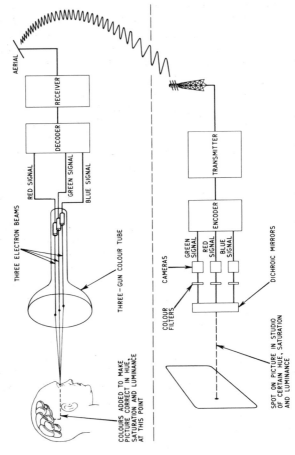

AERIAL

RECEIVER

DECODER

RED SIGNAL

GREEN SIGNAL

BLUE SIGNAL

THREE ELECTRON BEAMS

THREE-GUN COLOUR TUBE

COLOURS ADDED TO MAKE
PICTURE CORRECT IN HUE,
SATURATION AND LUMINANCE
AT THIS POINT

TRANSMITTER

ENCODER

GREEN SIGNAL

RED SIGNAL

BLUE SIGNAL

CAMERAS

COLOUR FILTERS

DICHROIC MIRRORS

SPOT ON PICTURE IN STUDIO
OF CERTAIN HUE, SATURATION
AND LUMINANCE

Figure 7.3. Instantaneous reception of a colour element using a three-gun tube

effective origin of the green beam, so far as the shadowmask is concerned, is the *deflection centre*, a point near the physical centre of the scan coil's field. It is here that the ultra-violet light source is placed during the phosphor-fixing stage in manufacture.

When the green phosphor has been selectively hardened in the required dot pattern, the shadowmask is removed and the screen is washed to remove the unexposed phosphor material which was in the shadow of the mask. We now have a uniform pattern of green-glowing phosphor dot 'islands' in the 'sea' of the glass inner-screen surface.

This process is repeated separately for the red- and blue-glowing phosphor materials with the ultra-violet light in these cases positioned in the deflection-centre point corresponding to the colour being fixed. At the end of the operation, then, we have a regular pattern or mosaic of dot triads aligned with the shadowmask and with the incident angle of the electron beams from the right guns. The pattern is shown in Figure 7.1.

There are several refinements carried out on the basic dot-mosaic screen for the sake of good contrast ratio and high efficiency. We saw in Chapter 2 that the darkest part of the picture can only be as dark as the unenergised screen itself, and the phosphor materials used are pigmented and dark-coloured to absorb ambient light falling on the screen and prevent reflectance. Additionally the small space between the phosphor dots is filled with a similar light-absorbing black pigment; these artifices increase the contrast range. Like a monochrome tube the shadowmask type is aluminised and this highly reflective screen backing increases brightness by reflecting phosphor-light forward instead of wasting it within the internal bowl cavity. The aluminium layer also acts as final anode and forms a protective ion-barrier for the screen.

Tube construction

Before we go on to examine the operation of the shadowmask tube let us look at some of its physical characteristics, starting at the back. The heaters are similar in concept to those used in monochrome tubes, but their power consumption is lower, and a quick-heat facility (less than five seconds warm-up) is achieved by concentrating the heating element just behind the front surface of the cathode. The neck diameter varies with tube type, but 29 mm is common in PIL tubes where a single 'gun' is used (still producing three beams of course!) whereas the older triple-gun Mullard 20AX and 30AX types have a 36.5 mm neck. An explanation of gun configurations will follow later.

Deflection angles are generally 90° or 110°. Very small colour tubes used in portable colour TVs have deflection angles down to around 50°, while at the other end of the scale some of the Sony Trinitron types have a deflection angle of 114°, which represents about the limit of current technology. The conductive inner and outer coatings of the 'bowl' section of the tube are resistive rather than highly conductive in many colour tube designs; these form an EHT reservoir capacitor with the glass wall of the tube as dielectric. The resistive coatings absorb suddenly-released energy in the event of internal flashover within the tube, and harmlessly dissipate it over a large surface area. Coming now to the shadowmask itself, we have seen that correct alignment of the beams on their own phosphor dots is essential for correct operation and good purity of colour. The correct alignment of the shadowmask is crucial to this. Because only about 23 per cent of the mask area is 'holes' and the rest is 'wall', most of the energy in the electron beams is dissipated within the mask rather than at the phosphors, and the total current flowing in the mask can approach 1 mA. With an EHT voltage in the region of 25 kV the dissipation in the shadowmask can thus exceed 20 W, and if this is sustained for long appreciable heating of the mask will take place. The resulting expansion must be taken up in a defined and controlled way if buckling and distortion are to be avoided – these would upset beam landing and purity. The mask, then, is either fitted under tension (Trinitron vertical aperture-grille) or in a special frame incorporating bi-metal support clips to permit an axial movement of the mask to take up expansion while maintaining hole-triad registration. As we shall see, Trinitron tubes use a mask consisting of continuous vertical slots, and this is more vulnerable to heating, especially local heating at a point of high brightness in the display. To give lateral support to the vertical grille members in large-screen Trinitron tubes, three very fine horizontal platinum tie-bars are micro-welded across the grille at intervals.

Regarding the screen itself, the size quoted used to be the diagonal measurement from corner to corner of the *glass faceplate*, larger than the picture. Today, screen sizes are quoted in V (visible) sizes of the *picture* diagonal. This may vary from 10 cm to 89 cm, though most domestic sets are fitted with 50, 56 or 66 cm tubes. The faceplate has two layers, the outer of which acts as an implosion guard in the event of sudden collapse of the glass envelope. These two front layers together are typically 2 cm thick in 66 cm tube, and this largely accounts for the weight of a colour receiver! The atmospheric pressure on the envelope of an evacuated tube is very great.

The PIL tube

Having set the stage, as it were, we shall now look at the basic operation of the shadowmask tube, initially the PIL type. PIL stands for 'precision-in-line' and refers to the configuration of the emitting cathodes, and the electron beams as they travel down the tube. The single electron gun of the PIL tube is illustrated in Figure 7.4, where it can be seen that a

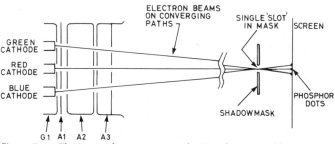

Figure 7.4. The internal arrangement of a PIL tube, viewed from above

121

common accelerating anode system and electron lens is shared by all three beams which travel through the gun assembly in very close proximity (5 mm) to each other. The three heated cathodes are in-line abreast at the left of the diagram, and their electron beams emerge through three small holes in the common control grid. They next encounter the first anode (A1) which accelerates them towards A2; the latter, in conjunction with the potentials on nearby A1 and A3 forms an electron lens for beam focusing. On emergence from the gun assembly the beams are on converging paths, as shown on the right of Figure 7.4.

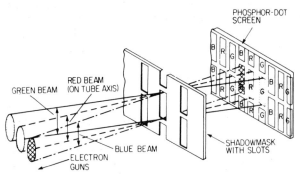

Figure 7.5. Working principle of the shadowmask and phosphor screen. For each beam, all but the intended phosphor dot is in the shadow of the mask

In fact the beams' trajectory is such that they cross over in the shadowmask hole and diverge just beyond to strike their appointed phosphor dots – an idea of this is conveyed in Figure 7.5, where the centre beam (red in our example) is on the axis of the tube with the outer beams, green and blue, having an approach angle of around 1° with respect to red.

Purity

Clearly, the approach angle of the three beams to the mask must be spot-on to get the 'shadowing' effect right. If the effective origin of the beams does not exactly coincide with the position of the light source used for dot fixing during manufacture, one or more of the beams will spill over onto part of a neighbouring phosphor dot which will be the 'wrong' colour for that beam. If this happens to red, say, and it spills onto the blue dots the red gun will give rise to some blue light from the screen and a magenta 'stain' will mar what should be a pure red field. To prevent this a pair of 2-pole ring magnets are provided around the tube neck in the gun area, and adjusted to achieve an on-axis beam path for good purity at screen centre. Secondly, the entire scan yoke is set axially (by sliding it along the tube neck) to establish coincidence between deflection centre and original dot-fixing light-source position. When this is correct, purity is established over the entire screen area for all three colours.

We have now achieved our aim, which is the establishment of three superimposed images, one in each primary colour and independently controlled for brightness. The triple beam may be regarded as a tri-coloured light pencil, with each colour winking and flashing as the picture is scanned out by the common deflection fields to describe a full colour picture. There is another factor to be considered, however, and this is the necessity to *register* the three images so that they exactly overlay at all points on the display; this is called *convergence*.

Converging the beams

What Figure 7.5 does not convey is that each beam is larger than one shadowmask hole, and in fact overlaps the slot edges. This means that not only must the beam approach angle be correct, but the spatial position of each beam must be adjustable to take up manufacturing tolerances and ensure that the three rasters are exactly superimposed; any error here will cause colour fringes on borders and outlines in the reproduced picture. What is required, then, is a means of independantly shifting the beams in the tube neck, and this is arranged by further ring magnets around the tube neck to provide well-defined and adjustable magnetic fields embracing *individual* beam paths within the electron gun. These fields are set up by a combination of four-pole and six-pole ring magnets.

There are a *pair* of ring magnets to set up the four-pole field, and a further pair to provide the six-pole field. Two are provided in each case so that by relative- and co-adjustment the *strength and direction* of the resulting magnetic field can be manually set to give complete control of beam positioning. None of these fields affect the central beam which travels straight down the axis of the tube after the fashion of that of a monochrome tube – this implies that the rasters due to the outer beams (green and blue in our case) are made to conform to the position of the raster traced out by the centre (on-axis) beam. Let us look first at the effect of the four-pole field. It is shown in Figure 7.6 and can be seen to move the

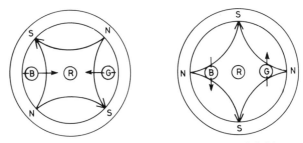

Figure 7.6. Four-pole static convergence rings with field pattern

outer beams *differentially* in either vertical or horizontal planes; the direction of shift being determined by field direction, and the amount of shift by field strength. This permits us to register the blue and green images to give a cyan overlaid image, but this will not necessarily coincide with the red image, and this is where the six-pole field comes into play.

The six-pole field is shown in Figure 7.7. Here the positions of the outer beams are moved together in the same direction, which may be vertical or horizontal in compliance with the orientation of the lines of flux. Again the flux strength (and hence the distance the beam moves) is controlled by the relative movement of the ring magnets, while the flux direction is varied by moving the six-pole rings *together*. Thus we can position our cyan raster exactly on top of the existing (and unvarying!) red raster.

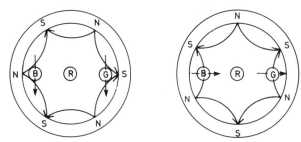

Figure 7.7. Six-pole static convergence rings with field pattern

On plain mixed-colour fields misregistration of the colours is not visible, so that convergence adjustment is carried out by observing the display of a crosshatch or dot-matrix pattern from a special signal generator.

Dynamic convergence

The convergence process described so far will only overlay the images in an area corresponding to a small circle at the centre of the screen. As the beams are deflected to left and right in the line-scanning process, the fact that the outer guns are off the tube axis means that the point of convergence will fall progressively short of the screen unless the latter is cylindrical with cylinder and deflection centres coincident as represented in Figure 7.8*a*. Because practical TV screens are much flatter than this, the three rasters become horizontally displaced (Figure 7.8*b*) and arrangements must be made to compensate for this. While a dynamic convergence system could be devised along the lines of the static magnets already

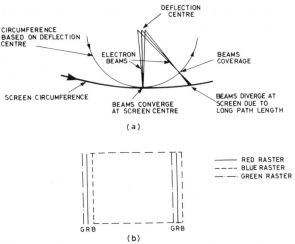

Figure 7.8. The causes of misconvergence: (a) shows how the beams diverge in areas away from screen centre, while (b) indicates the resulting misregistration

described, but with line-rate flux variations (i.e. a strap-on convergence yoke fed from the line timebase), a much neater system has been evolved, based on the concept of a special deflection yoke whose magnetic field is tailored to achieve correct dynamic convergence.

PIL deflection yoke

The degree of deflection suffered by an electron beam under the influence of a magnetic field depends on the strength of that field, i.e. how many lines of force it cuts. In a homogenous (uniform) field, all three beams will be 'bent' through the same angle as in our Figure 7.8a. Because the beams are horizontally separated by a few millimetres on their journey through the deflection field, however, we can arrange for them to encounter different flux levels by carefully 'grading' the magnetic deflection field across the tube-neck diameter; see Figure 7.9. Here the lines of force are widely separated in the area of the tube axis, and become progressively closer towards the tube's outer walls. This means that each of the outer beams will, during horizontal deflection, be turned into an area of stronger or weaker flux than its fellows, and so suffer a greater or lesser deflecting force. Provided the flux gradient across the tube cross-section is just right, complete convergence of the rasters can be achieved.

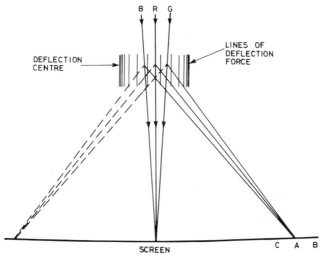

Figure 7.9. Beam trajectories in the deflection field. The latter is non-uniform to achieve correct convergence, as explained in the text

This effect is clarified by the beam trajectories in Figure 7.9. This is drawn from above and taking the case (solid line) of deflection to the right, we can see that the central red beam passes through an angle determined largely by the magnetic flux at the centre of the field to strike the fluorescent screen at point A. The blue beam, if it were similarly deflected, would cross the red beam some way short of the shadowmask and land at imaginary point B causing miscon-vergence. However, the blue beam can be seen to be (at the deflection point) turning into a weaker field, causing it to be deflected through a lesser angle than red and 'aiming' it precisely at screen point A. Now consider the green beam. In a uniform deflection field it would cross the red beam too early and diverge to strike the screen around point C. This time, then, we need to increase the green beam's deflection angle to achieve screen registration; as the green beam passes through the deflection field it is turned into a stronger deflecting field to achieve this. For line-scanning deflection to the left the opposite happens; in all cases the beam on the inside of the deflection-centre 'bend' is bent through a greater angle than the norm, and the beam on the outside is bent through a lesser angle.

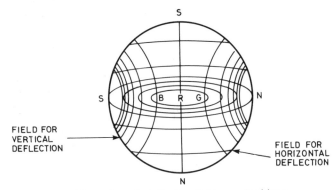

Figure 7.10. The astigmatic deflection fields required for a self-converging in-line picture tube. The field shapes shown are achieved by precision construction of the deflection yoke

What happens during vertical deflection? Again the yoke-to-screen beam-path gets progressively longer as we move upwards or downwards from screen centre, so that *horizontal* displacement of the three images will take place with increasing severity towards screen top and bottom. Again an astigmatic deflection field will correct this in similar fashion to that described for horizontal dynamic convergence. The vertical deflection field is barrel-shaped (Figure 7.10) so that the required horizontal lines of force have an increasing vertical component away from the tube axis. This vertical component has the effect of reducing the *horizontal* divergence of the outer beams to achieve horizontal register at screen top and bottom.

Thus correct convergence is maintained all over the screen area by a precisely-tailored deflection yoke. The positioning of the yoke along the tube axis is dictated, as we have seen, by the requirements of purity setting, and the gun end of the yoke is concentric with the tube neck due to their common diameter. To take up manufacturing tolerances, however, the flared front end of the scan yoke assembly may be tilted in both vertical and horizontal planes to 'trim' the relative trajectories of the electron beams and thus form a final adjustment of screen-edge convergence; this adjustment is carried out during manufacture or as a servicing routine, after which the yoke is wedged and sealed in position.

The necessary very precise field shaping is achieved in the PIL tube by a toroidal deflection yoke in which the positioning and distribution of the individual turns of wire are defined by notches cast in the supporting frame. The fields are further modified by magnetic 'shunts' and 'enhancers' fitted inside the tube or the scan yoke which respectively disperse or concentrate local magnetic flux to create exactly the field shape required for self-convergence.

20AX system

An alternative approach to in-line shadowmask tube technology is the Mullard/Philips 20AX system. The screen and shadowmask of 20AX is similar in concept and construction to that already described for the PIL tube, and while PIL types come in various sizes and have deflection angles of 90° and 110°, 20AX is confined to 56 and 66 cm screen sizes and 110° deflection. Instead of a single gun shooting three beams, this tube contains three separate guns, each very similar to that of a monochrome tube. The centre gun (green) is on the tube axis and the outer guns, for red and blue, are tilted inwards so that all three beams cross over within the shadowmask holes at screen centre. The construction of the 20AX gun assembly is shown in Figure 7.11.

The ring magnet system is very similar to that already described for PIL, with four- and six-pole magnet rings to set static convergence and a two-pole pair for purity adjustment. 20AX also uses an additional two-pole ring pair for geometric raster correction (symmetry), which is adjusted to eliminate bowing of the horizontal centre line.

Beams converge at slot-mask

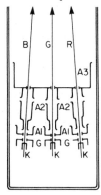

Figure 7.11. The 20AX gun assembly. Three separate electron guns send their beams on a converging path

The yoke of the 20AX system achieves the astigmatic field pattern described earlier for PIL tubes by a saddle-shaped winding pattern whose appearance is broadly similar to that of an ordinary monochrome deflection yoke. However, the wires are grouped into 'streams' to create the right field pattern, and the tube-internal magnetic components are not present in a 20AX type. Very close-tolerance manufacturing techniques in both yoke and tube eliminate the need for yoke manipulation in dynamic convergence setting. The yoke remains firmly on the axis of the tube neck, and any 'trimming' adjustments for dynamic convergence are catered for by a simple four-pole convergence correction coil associated with the deflection yoke. This carries sawtooth currents at line and field rate, whose amplitude and direction

130

are controlled by two or three preset controls or links; further convergence correction is made by differential adjustment of the currents in the symmetrical halves of each section (vertical and horizontal) of the deflection coils themselves.

30AX

Based on the technology of the 20AX system and the manufacturing experience gained, further developments led to the release of the 30AX display system, in which all the tube-neck adjustments were eliminated. These tubes are very similar in concept to the 20AX type, sharing the triple-gun and thick-neck characteristics and retaining 110° wide-angle deflection, but having a wider range of sizes, e.g. 50, 56 and 66 cm.

The 30AX gun assembly is longer than that of the 20AX because it has to accommodate a weaker electron-lens system – this gives improved overall focus. The 30AX has no neck-mounted magnetic rings; their place is taken by a small magnetic wire ring at the top of the gun assembly, into which is printed, at the time of manufacture, a composite magnetic field to fully correct convergence, purity and geometry. For a given tube size 30AX deflection yokes are fully interchangeable.

45 AX

A later Mullard/Philips design, designated 45AX, has a *unitised* gun similar to those described above in the PIL section. Here all electrodes downstream of the cathode are 'commoned' for all three beams, whose trajectories are governed together by the deflection system; and individually (for static convergence, geometry and purity purposes) by the single integral magnetic ring system pioneered in the 30AX. The main development here, however, is in the mounting of the shadowmask on a light diaphragm with a special suspension system to accommodate expansion of the mask with electron-beam heating effects. This greatly reduces the risk of purity errors.

FST

Current designs of TV receivers use FST (flat, square tube) designs in which the faceplate has a radius of curvature about twice that of a conventional tube, with straighter sides and squarer corners. The effect is to reduce ambient light reflections and give a more pleasing appearance to the set.

The Trinitron

The third group of in-line tube designs is represented by the Trinitron type which in fact was the first in the field, having been introduced by Sony in 1969. The main difference between this and the types so far described is in the form of the shadowmask. Instead of the discrete phosphor-dot triads of the conventional tubes, the screen of the Trinitron is composed of several hundred vertical stripes of red-, green- and blue-emitting phosphors. The mask consists of a metal

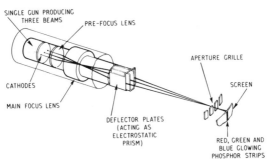

Figure 7.12. The main features of the Trinitron single-gun, three-beam aperture-grille picture tube

aperture grille which has one continuous vertical slot for each three (R, G, B) phosphor stripes. The general principles of the design are illustrated in Figure 7.12. It can be seen that (unusually amongst colour tubes) the beams cross over within the single gun assembly before crossing again at the aperture-grille slot.

The RGB beam crossover within the focus field confers the

advantage on this design of having all three beams pass through the centre of a relatively large electron lens. The analogy to light beams in optical lenses is obvious, and focal aberrations in the 'outer' beams are thus minimised. A further 'electron-optic' function is carried out further downstream by the deflector-plate system shown in the middle of Figure 7.12. These plates act as a prism due to a potential difference between them of a few hundred volts. By adjusting this p.d. the outer beams (red and blue) can be shifted laterally in opposite directions to converge them with the central green beam. The static convergence control, then, is associated with the EHT voltage source and connects to the prism plates via a coaxial EHT cable.

In other respects the Trinitron is similar to the PIL tube already described. Vertical static convergence is catered for by adjustable neck-mounted permanent-magnet systems, and such dynamic convergence trimming as is required is carried out by tilting the scan yoke. As in 20AX, a simple four-pole dynamic convergence coil is present and if necessary (depending on manufacturing tolerances) energised by a line-rate (and in large-screen types, a field-rate) adjustable sawtooth current derived from the timebase section.

Two advantages which the Trinitron offers over the opposition are the increased mask 'transparency', leading to higher picture brightness, and the enhanced vertical resolution, both due to the absence of a vertical matrix component in the shadowmask design.

Degaussing

Any external magnetic field will upset the beam path in a picture tube as a result of which beam landing accuracy – and hence purity – will be affected. Even the vertical component of the Earth's magnetic field can have an influence (the horizontal component cannot upset operation of an in-line gun tube) and nearby magnetic objects such as loudspeakers, radiators and steel joists can give rise to sufficient field strength to spoil the display.

133

To prevent these effects, the flare of the tube has a built-in magnetic shield, and to neutralise any magnetic field it may itself collect, a built-in degaussing system is provided in the receiver. It takes the form of a pair of coils embracing the tube flare (clearly visible in Figure 8.3) which are briefly energised at switch-on by a decaying burst of 50 Hz mains power. This comes from a double PTC (positive-temperature-coefficient) device whose resistance rapidly increases with the heating effect of the mains current so that the degauss energy decays to zero after two or three seconds, and remains so due to the presence of a 'bleed' resistor which keeps the PTC combination warm. To prevent pickup and distribution of line scan-yoke energy the degauss coils are bypassed at TV line frequency by a suitable capacitor.

Geometric distortion

In a monochrome picture tube, especially a wide-angle type, the flatness of the screen leads to a form of raster distortion known as pincushion, in which the raster borders are bowed in. This is corrected by permanent magnets which neutralise the effect by applying an equal and opposite pre-distortion of the scanning field. Colour tubes have similarly flat faces and so the same inherent defect is present, but any permanent-magnet correction system is not practical for reasons which should now be obvious! Instead the pincushion correction is carried out dynamically, as it were, by amplitude-modulating the line-scanning current with a field rate *parabolic* waveform, and by superimposing a line-rate sawtooth wave on the field-scanning current. This 'energy-interchange' is carried out by a transductor (saturable transformer) in some TVs. By this means the raster is pulled into a square shape.

The degree of pincushion distortion which has to be catered for depends on the type and vintage of the tube in question; early PIL types needed considerable correction on both axes, whereas later designs eliminated the need for vertical correction and reduced the horizontal requirement. Current PIL tubes, by further refinement of the yoke design, eliminate the need for pincushion correction altogether, and are called 'pin-free'. 20AX and 30AX require no vertical pincushion correction and 12 and 8 per cent respectively in the horizontal direction. This latter is arranged by a *diode modulator* in the line-scan stage which alters its tuning, and hence scan amplitude, on a field-rate basis.

Grey-scale tracking

A very important requirement is for the beam current versus cathode voltage characteristic of the three guns to match over the full range of brightness. Since it is impossible for guns to be manufactured with these parameters exactly matching (and to expect them to wear at equal rates), the tube has to be equipped with various adjustments to provide accurate *grey-scale tracking* as it is called.

We have seen that white light is produced when the mix of red, green and blue lights is correctly proportioned. This must occur at all drive levels from black to full brightness for each gun, for a drift out of alignment as drive is changed would alter the relative intensities of the three colours and thus change the proportions of the mixture, which would result in the white (or grey) display becoming coloured. The effect on a colour scene would be to alter the hue or saturation with changing luminance level. The black cut-off points are matched by varying the clamp potential on the individual cathodes by manual or automatic means, as we saw in the previous chapter. To match the brightness of the primary colours at high drive levels, the 'slope' of the video amplifier signal is adjusted by drive controls ('primary-colour contrast pre-sets') with which the highlights of a monochrome display are set to eliminate any coloured tint.

Focusing

We have talked much in this chapter of converging the beams for the purposes of picture registration, and this may be regarded as a focusing of the three beams inasmuch as we aim to get them to cross over at the shadowmask holes. Each individual beam may be likened to that in a monochrome tube, however, and the necessity to focus the beam itself is as important as in a black-and-white set; if good focus is not achieved, the colour image, while remaining converged, will have poor definition and render a blurred picture.

Each electron beam consists of a multitude of negative 'charge packets' – electrons – jostling and falling over each other in their headlong dash towards the highly-positive final anode. The fact that they are all *negative* charges makes them tend to repel one another, 'spreading' the beam. To correct this they pass through an electrostatic lens system (Figures 7.4, 7.11 and 7.12) in which the *outer* electrons in *each* beam are sent on an inward trajectory such that all the electrons in the beam come to a sharp focus at the phosphor screen. Adjustment of the focal length of this electron lens is carried out by varying the voltage on the focus anode(s) – usually A2; A3 in Trinitron types – by means of a potentiometer associated with the diode-split EHT transformer in modern sets, or the EHT tripler in older types. Typically, the potential required is about 18 per cent of final anode voltage in the conventional *bi-potential* tube type, and around 27 per cent of final anode voltage in the more modern *high bi-potential* tubes. With a 25 kV EHT voltage in both cases, this calls for around 4.5 kV and 6.8 kV respectively, with a focus control range of about ± 15 per cent. This potential is applied to one of the CRT base pins, and to avoid insulation and sparking problems the adjacent pins are widely spaced from it.

Beam current limiting

In our earlier discussion on the shadowmask, mention was made of the heating effect of the beam current flowing in the mask, and its special expansion arrangements to compensate. This works up to a limit, but at around the 1.2 mA mark (depending on screen size) the mask is at the end of its tether, metaphorically and literally! Beyond this point, purity suffers and convergence and focus become degraded. The *average* beam current in the tube can be measured indirectly by a monitor in the 'cold' side of the EHT supply, or directly at the cathodes. When the beam current reaches the upper limit, tube drive is automatically pulled back to prevent picture distortion. This 'pull-back' effect is carried out via the decoder (or luminance) IC by reducing the voltage available to the contrast, brightness and colour control-potential inputs.

Colour-tube performance

We have seen that there is one hole in the shadowmask per dot-triad, and because a TV pixel can be no smaller than a single triad, the definition capability of the tube is determined by the number of triads on the screen surface, assuming all other conditions of scanning, focus and convergence are optimum. In general the mask and dot structure are the limiting factor for resolution overall in a receiver, especially in the smaller screen sizes where the mask/hole matrix is not reduced in size in proportion to the screen area. Current medium- and large- screen direct viewing tubes are adequate for 625/50 colour reception, where the limitations on broadcast bandwidth offer a picture definition which can just be resolved by these tubes.

For higher-definition pictures and particularly high-resolution data and graphics displays, fine-matrix tubes are appearing with more triads per square cm, leading to an inherently better resolution capability. This puts further demands on the convergence system, which in any tube should ideally converge all three rasters to within one pixel over virtually all the screen area.

Another changing feature is the shape of the viewing screen. The new generation tubes have squarer corners and straighter sides than conventional types, and conform to the ideal 'rectangle' shape with an aspect ratio (width-to-height) of 4:3. This is the logical conclusion of the trend over the years, starting with a circular tube surrounded by a quasi-rectangular mask.

Projection TV

For a large picture the direct-viewing system we have been describing is not practical for many reasons, foremost of which are weight, bulk and the effect of atmospheric pressure on a large evacuated glass envelope! Until a practical large flat-screen device is developed (no easy matter, as we shall see) the only alternative to direct-viewing CRT displays is some form of projection system. The two basic approaches to this are to project the coloured image from a conventional-format tube onto a large screen by means of a lens; or to project separate RGB images onto a common viewing screen, rather like the set-up we used in Chapter 3 to illustrate the basic concept of colour TV and light mixing. The former approach has been used in domestic applications, but its light output is pitifully low since the limited CRT phosphor-light becomes spread over a large area. RGB systems are more complex and expensive, but are capable of much brighter picture reproduction for reasons which will become clear, and it is on these that we shall concentrate now.

In a projection system, many configurations are possible, and much depends on the required screen size and viewing area. To make a (relatively) compact set, the light path is often folded by an intermediate mirror and the image formed on an integral screen. This system may use *back-projection*, where the picture is built up on the rear of a ground-glass screen which (for domestic and small-audience commercial use) is fitted into a free-standing cabinet which thus resembles a huge conventional console TV set. An alternative and more common approach is the front-projection system where projector and viewers are on the same side of a reflective screen. This offers the great advantage of being able to give the screen a 'gain' factor by building a directional property into it – the concave screen throws a lot of light over a small viewing area and a useful increase in brightness is realised in the limited viewing area. Off-beam the brightness is very low indeed. A light 'gain' of three to five times is possible with a directional screen.

In the sort of RGB projection system we are discussing here, the light sources are three separate picture tubes with different phosphor screens, one each of R, G and B-emitting – corresponding to the three TV primary colours. The chief requirement is high brightness and with no shadowmask to intercept beam energy, and a high accelerating (EHT) voltage, the limiting factor for brightness is the heating effect of the beam dissipation on the glass faceplate. Very expensive projection tubes are made with sapphire faceplates to withstand high temperatures, but a more homely arrangement is the use of a forced-air cooling system for three conventional small cathode ray tubes. These may be mounted in-line abreast like the guns of a modern picture tube, or in delta formation analogous to the gun arrangement in early shadowmask picture tubes.

The next operation is to gather as much light as possible from each tube and project it onto the viewing screen, along a straight or folded light path. Either conventional lenses or a *Schmidt* optical system can be used; the basis of the Schmidt system is a highly-polished spherical mirror which gathers the scattering light rays from the tube and concentrates them into a parallel beam, the reverse process to that used in astronomical telescopes, or indeed the satellite receiving dish to be described in the next chapter. The parallel beam passes through a corrector lens on its way to the viewing screen. To prevent contamination of the highly-polished reflector and to ensure perfect and permanent positioning, the entire Schmidt system is built into the tube envelope in some designs – Figure 7.13 represents the LightGuide tube by Advent Corp of the USA. The internal screen is about 9 mm diagonal, and made of aluminium, coated with R, G or B phosphor.

Whether simple lenses or Schmidt units are used, the fact that the three projection lights are not coincident in position means that the light beams strike the viewing screen at different angles and, as with our shadowmask picture tubes, this causes misregistration of the three primary colour images. To correct this, a convergence system is incorporated in the deflection system of two or more tubes to

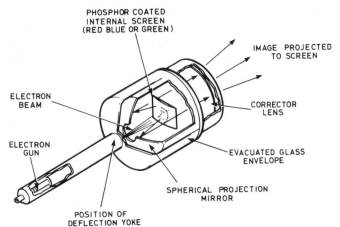

PHOSPHOR COATED
INTERNAL SCREEN
(RED BLUE OR GREEN)

IMAGE PROJECTED
TO SCREEN

ELECTRON
BEAM

CORRECTOR
LENS

ELECTRON
GUN

EVACUATED GLASS
ENVELOPE

SPHERICAL PROJECTION
MIRROR

POSITION OF
DEFLECTION YOKE

Figure 7.13. An internal-Schmidt projection tube (Lightguide) by Advent Corp of USA

introduce an equal-and-opposite distortion to each individual raster, and these are finely adjusted to overlay the three images on the viewing screen. Where the screen (or the incident light beams) is at an angle to 'normal', keystone distortion will also be present, and this is corrected in the same way.

Light valves

In all the systems discussed above for TV display, the basic source of illumination is phosphor-light, and for very bright or very large displays this technology is pushed to the limit. If we were able to emulate ciné practice, and use the picture information to *modulate* a separate, external light source on its way to the viewing screen, overall illumination would be limited only by the brightness of the 'lighthouse'. This is not easy to arrange for TV and is certainly not suitable for domestic use. The best-established system uses the Eidophor principle, wherein the TV raster is traced out on a spherical mirror in an evacuated glass envelope. The face of the mirror is coated with a special oil whose surface is deformed by the electron beam. The TV lines are formed as grooves in the oil film, and the depth and wall-angle of these grooves depends on beam intensity – picture brightness. A carefully-formed and precisely-angled light is focused onto the oil film which then passes or blocks the light so that the spherical mirror's reflectance is controlled by the TV signal. Emerging light is focused on the viewing screen by a lens, and the whole shooting-match is triplicated for the three primaries of colour TV.

An alternative light-valve system, developed in France and called the *Titus* tube, uses a single crystal plate of deuterated di-hydrogen potassium phosphate ($KD_2 PO_4$) which is bombarded from the rear by a constant-current electron beam (describing the lines of a plain, unmodulated television field) and from the front by a light source. The latter is reflected by a dielectric mirror at the rear face of the crystal plate. The incident and emerging light beams pass through crossed polarisers and the effective reflectance of the plate/mirror combination is dependent on the *birefringence* of the crystal plate at each pixel. This is varied by the electrostatic charge between the plate and an adjacent rear-mounted mesh-grid. The electrostatic charge pattern is set up by direct connection of the video signal (about 125 V black-to-white) between the plate and mesh, and removed by the scanning beam as it picks-off each pixel in turn, a process reminiscent of the vidicon camera tube.

Other light-valve TV display systems are currently under development and show promise, but it is unlikely that they will be economically viable for domestic use.

Flat-screen colour

A great deal of research is currently going on in the area of flat-screen displays and while monochrome systems have appeared in several (mostly low-definition) forms involving conventional phosphor screens, LCD (liquid-crystal display) and even matrices of ordinary lamps (Mitsubishi Diamond Vision, an outdoor 6×8.5 m analogue display), the difficulties of providing three colours with sufficient brightness, and an adequate number of pixels, have proved very great. Some small-screen colour LCD displays are used in 'pocket' TV sets, albeit in low-definition form. A rather complex flat screen display system was developed by RCA of America, using phosphor dots, a shadowmask and electron guns; study of this is instructive . . .

The arrangement is shown in Figure 7.14, and consists of an evacuated rectangular box 10 cm thick with a screen diagonal of 125 cm. The glass screen surface is completely flat and has an internal mosaic of R, G and B phosphor dots backed by a shadowmask in very similar fashion to a conventional tube.

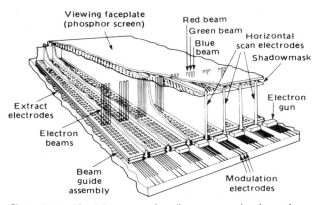

Figure 7.14. The RCA approach to flat-screen technology: the guided beam display system

The electron guns and associated modulators are mounted at the bottom of the panel so that their electron beams (one for each colour per horizontal pixel) travel up the panel parallel to the plane of the screen. The path of each electron beam takes it up a *ladder guide* which is perforated to match the shadowmask. This does for the electron beam what a fibre-optic cable does for a light beam, and its transmission efficiency is very high.

Unlike a fibre cable, however, the ladder guide is designed so that the electron beam can be diverted through a right-angle to emerge through any of its holes to strike the

142

phosphor dots – after 'filtering' by the shadowmask of course – and thus produce a colour picture. This sudden 'exit stage centre' process is triggered by the application of a negative potential to the appropriate horizontal extract electrode on the rear wall (bottom-most layer in Figure 7.14). The voltage required is -60 to $-180\,V$ which instantaneously repels the electron beam towards the phosphor screen – a form of electrostatic deflection.

There is one extract electrode strip per vertical pixel, so that picture resolution is directly geared to the number of ladder guides and extract electrodes fitted. The system works in digital mode with both scanning and video drive, and is incapable (even in its most developed form) of approaching the brightness performance of a conventional tube. Resolution performance is comparable, and the guided-beam display does not suffer from the loss of resolution at high brightness levels inherent in the conventional type. The aesthetic advantages of this particular system have a trade-off in complexity of drive circuits; there are around 400 electrical connections to the display panel and a great deal of peripheral circuitry, including a high-voltage oscillator for horizontal scanning and RAM memories to store the emission characteristics of the electron guns. Study of this sort of display technology brings home the realisation that the conventional shadowmask tube is an incredibly simple, efficient and effective device, with only 12 electrical connections required for its operation, including scanning.

8

The colour receiver

A colour TV set can be regarded as a monochrome receiver with refinements, to which has been added the colour display device and a decoder. Figure 8.1 shows the block diagram of a colour receiver with the 'extra' items required for colour in heavy line.

The colour display device is shown in heavy line because this is specific to a colour receiver, but as we have seen, a colour set can display very satisfactory monochrome pictures, whether or not the transmission is in colour. Thus the main and essential item in the colour set, apart from the display device itself, is the decoder, which has been fully dealt with in Chapter 6. In most other respects, monochrome and colour receivers are similar or identical – each contains a tuner with some means of viewer control; an IF amplifier with carefully-shaped frequency response; a vision detector to produce video and sound signals; an 'intercarrier' audio IF and detector stage; one or more audio amplifiers and loudspeakers; a sync-pulse separator; line and field timebases; and a power supply unit. The degree of sophistication of many of these circuit blocks needs to be greater for colour TV, and much depends on the price of the model in question, especially in such areas as remote-control, sound quality and interfacing ability. Many colour TV receivers are fitted with Teletext facilities, a feature not available on any production monochrome set! In recent years the production volume of black-and-white sets has been very low, consisting in the main of portable, small-screen types.

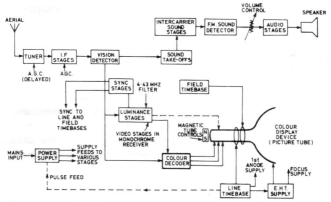

Figure 8.1. Block diagram of colour receiver. The heavy-line block represent the extra stages required for colour

The function and operation of the various 'common' circuit sections are well known, and do not merit a full and detailed description here. Other reference books (e.g. *TV and Video Engineer's Pocket Book* by the same author) cover them in detail. For the sake of completeness, however, brief descriptions are given below.

The tuner

The tuner's job is to amplify the weak signals picked up by the aerial, to select or 'tune' the required one, then to amplify and translate it, complete with all the information that it carries, to a lower (intermediate) frequency called the 'IF'. It will be recalled that the aerial itself is a tuned circuit, so the aerial and tuner should be regarded as a 'matched pair', a point perhaps which should be borne in mind by the aerial-erecting fraternity!

It should be remembered that the selected signals are in a specific channel and consist of both the modulated sound and vision carriers. Reasonable balance between all components of the transmitted signal on any one channel is

145

required, though FM sound limiting and the ACC function make the set reasonably tolerant of 'tilt' in the response of the aerial or tuner section. Most of the gain and selectivity of a TV receiver is designed into the IF amplifier, and the main requirements of the tuner are a low inherent noise figure and an ability to suppress the 'image' frequency, which the IF amplifier is quite unable to reject. Normally the local oscillator (these tuners are, of course, superhet devices) runs 'high' with respect to incoming frequency, and the IF output from the tuner is the *difference* between the two, normally 39.5 MHz for vision. There are two input frequencies which will give rise to this 'beat' frequency, one 39.5 MHz above the oscillator frequency, and one 39.5 MHz below. The former is the so-called 'image' and any signals arriving at the tuner on this wavelength must be strongly rejected.

The tuned circuits in the tuner unit must keep in step with each other as different frequencies are selected, and this calls for good matching in manufacture, or a precise alignment during production. The variable tuning is carried out by means of fixed Lecher bars (inductors), in conjunction with varicap diodes which we met in our discussion on colour decoders. Tuning control, then, is carried out by varying the DC potential applied to them. The control voltage source may be as simple as a single potentiometer or as complex as a frequency-synthesis, self-seeking ensemble. Perhaps these last terms deserve a little more explanation! A self-seeking system, when initiated, sweeps up the TV transmission band(s) by itself, stopping each time it encounters a station for committing (manually or automatically) to its solid-state memory. Frequency-synthesis tuning offers a self-seek and memory facility along with a 'direct addressing' feature in which a required channel *number* (21 to 68 for UHF) can be requested by the viewer and automatically tuned. This involves a very stable crystal oscillator in a PLL embracing the tuner's local oscillator. The PLL includes a *programmable divider* to set the ratio between crystal and local oscillator and this division ratio governs the tuning point of the TV. All this is achieved by special digital ICs.

146

IF amplifier

Looking again at Figure 8.1, the level of IF signals (both sound and vision) from the tuner are raised by the IF amplifier, which consists of one or two ICs. The response characteristics of the stages are carefully tailored to selectively amplify sound, chroma and vision signals, and reject out-of-band products, by a SAW (surface-acoustic-wave) filter, which may almost be regarded as a cross between a crystal and a glass delay line! With selectivity provided by this filter, and gain by the following amplifier, the IF signal is brought to a condition suitable for demodulation, or detection.

Vision AGC

Automatic gain control is applied to both the tuner RF amplifier and the IF amplifier. The vision signal is *sampled* (usually at line rate) to measure its amplitude, and the sample is produced in DC form as an *error* voltage whose level is proportional to signal strength. This is fed back to an early stage in a way that varies its gain, thus setting up a feedback loop which maintains a constant output level from the vision detector.

While the *level* of signal can be held constant by AGC over a tremendous range of RF input levels, the *noise* (snow/grain effect) is proportional to received signal strength. Below about 300 mV (for modern UHF tuners) this noise becomes obtrusive, and to achieve the lowest possible noise level in weak signals the gain of the tuner is held at maximum unless the signal becomes so strong (2–3 mV) that overloading of the tuner could occur, when its gain is automatically reduced. This is called *delayed* AGC.

Vision detection

The IF signal is demodulated within an IC by a synchronous detector working on a sampling system similar in essence to that employed for colour demodulation in a decoder. In place of the local oscillator we have a 'tank' circuit which is energised by the incoming IF signal itself, and critically tuned for linear demodulation. This is followed by a filter to remove residual IF carrier components. The vision signal is now ready for application to the decoder block as described in Chapter 6.

Intercarrier sound

At the vision detector input both vision (39.5 MHz) and FM sound (33.5 MHz) signals are present, and it is arranged that intermodulation takes place between these two. As a result, a *difference* frequency of 6 MHz is produced, containing modulation components of both contributing signals. Its AM (vision) modulation is minimised by a 'swamp' technique of depressing the 33.5 MHz sound carrier (in the early vision IF stages) to a level below that of the lowest (i.e. peak white) excursion of the vision signal. A tuned acceptor circuit picks off the post-detector 6 MHz beat signal for processing in an IC containing a series of amplifiers and limiters which remove any residual vision components, leaving a pure FM signal at 6 MHz containing sound information. After detection – again a synchronous, *quadrature*, detector, is used – the audio signal is back in baseband form, without any vision buzz if the limiters are doing their work! Amplification and processing in volume and tone controls render the audio signal in a form suitable for driving a loudspeaker.

In the MAC receivers (to be described in the next chapter) the audio signals, which as we know are in stereo, are produced in a totally different way. Here we have a strobed data-slicer which produces numbers corresponding to the quantised audio signal levels; they become recognisable after their passage through a D/A converter.

Sync separator

The detected video signal also feeds the sync separator to produce the all-important timing data for scan synchronisation, described in Chapter 1. This is a relatively simple circuit, internal to an IC, which discriminates between vision and sync signals on an amplitude basis, a sort of electronic bacon slicer, as it were. Usually a noise-cancelling circuit is provided in the chip which disables the separator during interference bursts, preventing picture judder and tearing. Often the sync separator is in the same package as the line oscillator, and many modern chips incorporate sync separator, line and field oscillators and several associated functions in a single IC package.

Timebases

The vertical and horizontal scans are produced respectively by the field and line timebases subjecting the electron beams in the picture tube to appropriate deflection forces by electromagnetic means, as explained in Chapters 1 and 7.

A substantial amount of magnetic force is needed to provide full deflection of the beams vertically and horizontally, and this force is generated in the field of the scanning coils on the neck of the tube. Figure 8.2 illustrates a typical deflection yoke and associated line output transformer for use with an in-line gun picture tube.

The line timebase (or the power-supply unit, see later) drives a special circuit for producing the accelerating (EHT) voltage for the picture tube. In current receivers this consists of a *diode-split* transformer section working at line frequency (15.625 Hz) wherein several separate EHT secondary windings are used, each forming a 'cell' in association with a rectifier diode and a reservoir formed by inter-winding layer capacitance. Each of the EHT cells generates, say, 5 to 8 kV, and they are connected in series to yield an EHT potential appropriate to the tube size in use. Except for very small screens, this will be between 20 and 28 kV.

The focus voltage for the picture tube will also be generated in this area by 'tapping off' the output of the first or second EHT cell for application to the *thick-film* focus control potentiometer. Likewise, the A1 potential and heater energy for the tube are sourced from the line-output transformer. These supplies – and in many designs other LT

149

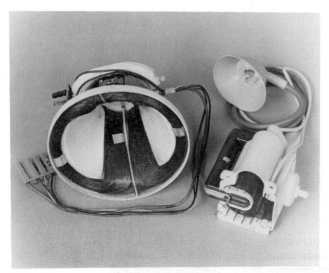

Figure 8.2. Accessories for use with a 90° PIL tube: the deflection yoke and diode-split line output transformer

and HT lines – have good *regulation*, benefiting from the close stabilisation which must necessarily be applied to the operation of the line output stage to ensure constant scan amplitudes in the face of changing EHT current demand (with varying brightness) and mains voltage fluctuations.

Power supply

There are two approaches to power-supply design in current practice. The first depends on a mains-driven *primary* power supply which generates primarily an HT line of between 100 and 220 volts, closely stabilised, to operate the line-output stage from which all or most secondary supplies for the rest of the receiver are derived, as explained above. The alternative method involves a fairly simple line output stage driving the scan coils and EHT generator perhaps, while all the auxiliary supplies come from the PSU itself by means of separate rectifier/filter sets. Some designs also incorporate the diode-split assembly in the PSU or *chopper* transformer. In either case the PSU works at high frequency and high efficiency, as will be briefly described now.

In a TV set the less energy dissipated the cooler the set will run, making for greater reliability and lower running costs. To achieve an efficient stabilised power supply at reasonable

Figure 8.3 An internal view of a TV receiver using the 30AX picture tube. This is the Ferguson TX10 chassis, with baseband video and audio inputs (top right) and facilities for stereo sound and Teletext, the circuit boards for which are on the left-hand side of the chassis (photo courtesy Ferguson Ltd)

cost the 'chopper' technique is used, in which the regulating element is an electronic switch. The incoming mains is rectified and smoothed in a mains-live area of the set and applied via a high-speed switch to the primary winding of a ferrite-transformer. The switch is in the form of a transistor or thyristor working at a rate of several kHz, most often synchronised to line-scan rate, as shown by the dotted link

151

between LTB and PSU in Figure 8.1. The ratio of on-time to off-time of this switch (that is the mark-space ratio) is varied, thus regulating the total energy present in (and passing through) the transformer. Mark-space variation is carried out by a drive-pulse generator which monitors the output voltage from the PSU and adjusts its timing accordingly. Mains isolation of the set and its metalwork is achieved by insulating the primary and secondary windings of the 'chopper transformer'. This now-universal technique is very efficient and can cater for wide variations in mains input voltage and tube energy demand. This is reflected by the power consumption rating of the set itself which is typically 50–80 W at full picture brightness, and considerably less at lower brightness levels. A photograph of a typical colour TV receiver appears in Figure 8.3.

Advanced CTV

In the remainder of this chapter we shall examine some of the techniques, proposals and philosophy of colour TV which are currently under development or being suggested for TVs of the future. Improvements and innovations in home television take a long time to implement because of the 'inertia' factor of an established system and the existence of millions of TV sets working to it. In the last twenty-five years we have seen the change from 405-line to 625-line TV, the introduction of colour and videotex, and the growth of 'TV sources' from two monochrome broadcast channels to a multiplicity of programme material from four national colour channels, videocassette recorders, disc players, cable and space.

Innovations will continue and technology will make possible many improvements to the colour television system, as we have already seen in earlier chapters, in the chapter on display devices, and will see in the next two chapters. The limiting factor in implementing new TV systems with enhanced performance are bandwidth, which sets a limit on the detail which can be passed through a system (especially a broadcast channel) in unit time; and cost, not so much to the broadcasters and programme originators but to the consumer in terms of how deep he must dig into his already-committed pocket for his total installation. In the area of sound reproduction many consumers are prepared to pay dearly for good equipment – if this trend extends into the television field (and all television is a bargain in real terms!) we may see a great improvement in the quality of picture reproduction in the home.

Teletext and enhanced text

The basic principles of Teletext are described in *TV and Video Engineer's Pocket Book*. The system utilises 'spare' lines in the field blanking interval to carry a digital pulse stream which when decoded can type out on the TV screen data in the form of text and simple graphics (maps, diagrams etc.). At present 8 to 12 TV lines per field are used for this service, preceding line 18 in odd fields and line 331 in even fields. These data signals are visible as twinkling dots at the top of the picture if the scan height is reduced on a 'live' picture display.

The current Oracle and Ceefax text transmissions, while capable of up to 800 pages per magazine (per channel), contain 100 to 200 pages apiece to minimise access time. On receipt, the requested page is written into a digital memory for continual read-out and update on the screen. In *Fastext* systems associated pages are also acquired and stored for instant display at the touch of a colour-coded 'prompt' button on the cordless control keyboard. The amount of data available, or *bit-count*, determines the number of effective pixels in the data display, and hence its detail and resolution.

This bit-count depends on several factors. The capacity of the page store at the receiver determines how much detail can be held for display during one 20 ms field, although the actual characters and graphics blocks in that field come from a character-generator chip programmed from the page memory. The next limiting factor on bit-count is the time available for transmission of the data. Where data is slotted into the spare lines of a programme, only about two per cent of the available time is dedicated to data transfer, and this is further reduced by the need for *framing* or synchronisation pulses, error-correction pulses and the necessary 'secondary' data to control the colour, size and other characteristics of the data display. Finally comes the question of *bit-rate*. This depends on the bandwidth of the system, that is to say, how many noughts and ones we can get down the 'pipeline' in unit time. For Teletext, digital pulses are sent at 444 times line rate, giving rise to a maximum frequency of about 3.5 Mhz.

Even within the above constrictions the result is very acceptable. In the current text format we can display 23 rows of characters with a maximum of 40 characters (or spaces!) per row. In the graphics mode, a maximum of 6000 pixels are possible, but with severe limitations on colour selection (any change of colour along one row must take place at a 'gap' in the row due to the use of *serial attributable characters*). The current UK system of Teletext and the very similar Prestel (which comes via a BT telephone link) are called *Basic Alpha-Mosaic* and a typical text-and-graphics page is shown in Figure 8.4. While this is very acceptable for transmission of weather maps, simple diagrams and charts, they have necessarily to be drawn with a broad brush, as it were; certainly the possible total of 920 alpha-numerical characters per page permits a most adequate 'message' to be presented on a single page.

The basic alpha-mosaic system which we have at present can be greatly improved by getting more data into the page display. The CCIR (International Radio Consultative Committee) has proposed five levels of teletext, of which the present Teletext/Prestel system is the lowest. Level 2 is called *enhanced alpha-mosaic* and offers a more versatile character set to embrace most European languages, smoother mosaics, a facility to address individual text-pixels separately (any pixel may be any colour) and an increase in the range of colours available (level 1 offers only a choice of eight).

Level 3 offers DRCS (dynamically redefinable character sets) which are transmitted along with the magazine data. In effect the decoder's character generator is programmed by the incoming data, not only regarding which character to print, but also how to form each character. Thus a facility for displaying special characters and foreign alphabets to order is provided, along with better graphics, based on a 12 × 10 dot matrix in each character cell. At level 3 these facilities can be used for either high-resolution or multi-colour as required, and it is possible for instance to provide quite high-resolution colour graphics over a limited area of the display.

Figure 8.4. Level 1 Ceefax. This is reproduced from an off-screen colour photograph supplied by the BBC

With such systems as the Canadian *Telidon* (text level 4) we move out of the area of alpha-mosaic displays and into another mode, *alpha geometric* where typically 75 000 pixels are available in the display, each of which can be separately addressed and assigned any one of the available 16 colours. For any given picture these 16 colours can be chosen from a quarter of a million available! Display of level 4 text calls for a degree of computing power and a great deal of storage capacity at the receiver if the full potential of the system is to be realised. As will be appreciated, the method of encoding level 4 text is quite different to that for the alpha-mosaic systems, and each pixel is here defined on X and Y coordinates; the image is built up on the screen according to transmitted 'geometric' commands called *picture description instructions* such as point, line, arc, rectangle and polygon. An idea of the potential capabilities of level 4 can be gained from Figure 8.5 where the banana display contains 2600 bytes of data.

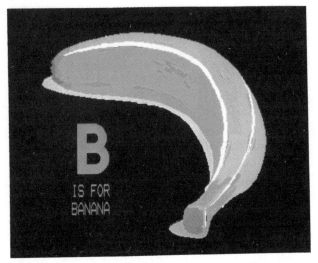

Figure 8.5. Level 4 graphics. Reproduced from an off-screen photograph supplied by the IBA

Finally, we come to text level 5. This is known as *alpha-photographic* mode and can, in its most developed form, offer a still-frame display better than that of any analogue-derived picture with a similar pixel count, which is limited only by the resolution of the display device, and the bandwidth of the transmission channel it occupies. As in level 4, each picture element is separately addressed, but now with 256 possible levels for each of R, G and B per pixel. This calls for the transmission and storage of a great deal of information, and would for instance require about 2½ minutes to build up if sent over the currently-available text channel in a normal broadcast. To store the information for display, a page store at the receiver would require a capacity of 1.2 M bytes (a byte is an eight-bit *word*) which amounts to a full field store. Fill this store at 40 ms intervals and read-out

Figure 8.6. Level 5 enhanced Teletext. A full-definition frame with the capability of reproducing high-quality full-colour stills (photo courtesy BBC)

at a suitable rate and we are back to moving pictures, digitally transmitted! Figure 8.6, which is reproduced from a BBC colour photo, indicates the resu ts obtained in a level 5 extended teletext experiment.

This hierarchy of text levels can be engineered for compatibility between generations of decoders, so that the display will be dependent on the sophistication of the decoder fitted to the viewer's set.

High-definition and high-fidelity TV

As with the text systems just discussed, the path to high-definition TV consists of several discrete steps, starting with improvement and enhancement of the existing display on current standards, and ending with the realisation of a TV picture with a wide-screen format and resolution capability equal to – or better than! – the best 35 mm cine film reproduction, and sound to match. As we shall see, the technical problems associated with high-definition TV are huge, but not insurmountable. Where home and broadcast TV are concerned, the old bogey of compatibility rears its head again, and many of the proposals being made are based on the idea than any new system will be amenable to existing receivers; an incompatible format would call for dual-standard receivers and possibly programme duplication, which is wasteful of spectrum space, no matter which media (or band) is used for transmission.

The digital MAC system, shortly to be described, is a first step towards enhancement of the TV picture. Although based on the traditional 625/50 system, the elimination of the worst visible defects of subcarrier colour systems such as cross-colour offers a great subjective improvement in repro-duction. By incorporating two digital field-stores in the receiver, further improvements are possible. By interpolating between corresponding lines in the two interlaced fields and drawing more lines in the display, it is possible to achieve a subjective definition equivalent to a 900-line picture. The presence of a frame-store also facilitates reading out picture information at a faster rate than the 50 fields/second used in normal transmission to eliminate the flicker effect which is noticeable in large areas of high brightness.

To achieve compatible high-definition TV, one possibility is to use one broadcast channel for a standard picture, and a second channel to provide an enhancement signal which would provide the extra detail (or even the 'sidepieces' of a wider CinemaScope style display) to fill out the basic picture. This concept is a little alarming from the point of view of tuning and balancing the two components of the enhanced picture, however, and conjures up the vision of sidepieces with no centre, or 'mid-fi' television with a half-cocked enhancement channel! The idea of 'basic colour TV' with an optional extra, as it were, was the basis of the BBC's E-PAL system described in the next chapter.

More than 625 lines

We have seen many times in this book that more picture detail calls for more bandwidth, and this is true whether the signal is transmitted in analogue or digital mode. For a worthwhile gain in picture definition we have to go beyond 1000 lines, and systems have been suggested using between 1100 and 1500 lines. A doubling of the European 625 standard to 1249 lines has often been advocated in the interests of compatibility and Philips have done much research along these lines. To draw 1249 lines in the same field period of 20 ms the scanning spot has to move twice as fast as before which means that to draw the same features the video signal must also move twice as fast, implying a doubling of bandwidth. Having done this, however, we are left with a picture which is capable of resolving much detail provided such detail is stacked vertically to take advantage of the higher line standard.

There is little point in doubling the vertical definition without a proportional increase in horizontal definition to match. To achieve this we need to double the vision signal bandwidth again, and so we arrive at a figure of 20–25 MHz for the luminance channel alone, or for each channel if the system is an RGB one. Such bandwidths are not easily accommodated, especially in air and space broadcast where even the 12 GHz satellite band is not a suitable medium. There are higher SHF bands (22 GHz and 40 GHz) suitable for satellite use in broadcasting high-definition TV, but the technology for implementing this is not fully developed at the time of writing.

159

Aspect ratio

The aspect ratio of current TV displays is 4:3, largely because the first picture tubes had circular screens and the TV standards were based on these. For greater realism, it is generally accepted that an aspect ratio approaching 2:1 is a reasonable compromise between the ideal (which presumably is a circular screen encompassing the viewer) and the practicalities of generating the display. Certainly the wide screen goes better with stereo sound, and Hi-Fi TV will incorporate both these features.

High-definition TV programme sources

While 'electronic' signal sources will provide data displays and animated cartoons (along the lines of the text systems discussed earlier in this chapter), most people are more interested in 'real' pictures, and these must come from cameras and ciné film. Solid-state image sensors have now been developed to the point where they could be used for picture pick-up in an HDTV system with the advantage of excellent RGB registration. Early experiments and demonstrations of HDTV necessarily used very-high definition camera *tubes*, based on the vidicon and Pumbicon principle, but usually of larger diameter, perhaps 30 mm.

The hardware for televising ciné film is called a *telecine* and most use not camera tubes but photomultipliers for image sensing, in conjunction with an FSS (flying-spot-scan) tube to illuminate, 'swinging torchbeam' style, the film frames as they pass the gate. At the very high data rate involved in high-definition TV neither photomultipliers nor FSS tubes are really fast enough, and experiments have involved CCD line-sensors in conjunction with a prism for pick-up (Philips) and a laser beam deflected by a rotating polygonal mirror as light source (NHK, Japan). It is fortunate indeed that good 35 mm (and especially 70 mm) ciné film contains sufficient detail to do justice to these new television systems!

Ironically, once high-definition TV is fully established in the professional closed circuit (as opposed to the broadast) field, it may well signal the end of ciné film for professional purposes. Provided the necessary parameters are met, electronic film-shooting has many advantages over a conventional film camera in terms of versatility in subsequent processing; creation of special effects; correction of lighting and other deficiencies in the original scene; lower production costs in scenery and backgrounds, which can be subsequently 'added' by electronic mixing; and the ability to instantly review a sequence on a VTR monitor before sets are dismantled and personnel dispersed.

High-definition displays

The shortcomings of the shadowmask tube have already been discussed in relation to conventional television, and we saw that the shadowmask/triad combination sets a strict limit on the resolution available from this type of display. A well-set-up and magnetically-focused monochrome tube can display 1200 lines without too much trouble, but to achieve this with a shadowmask system is beyond practical and economic reason. There is another factor, and that is screen area. Even if we could achieve an 1149-line picture on a colour tube of conventional size (50–70 cm), it would subjectively show little resolution improvement over the present format, unless the viewer were within a few centimetres of the screen. For home use, a display area about 1.4 m × 0.7 m seems optimum for high-resolution TV, and conventionally-scanned tubes are not very amenable, in a physical sense, to that size or shape!

To display our high-definition pictures, then, with sufficient brightness and resolution it is currently necessary to use projection techniques. For very large-screen applications one of the light-valve systems described earlier can be used, and the Eidophor system is demonstrably capable of handling pictures with more than 1000 lines. For smaller screens in the 1 m² region, direct-view tubes are now challenging projection systems in which the requirements of registering all three images within half a pixel all over the screen make tremendous demands on the convergence system.

Figure 8.7 shows an experimental Hi-Fi TV projector by Philips. It can define one million pixels, and uses three

Figure 8.7. A projector for Hi-Fi TV by Philips. The lens system for the R, G and B tubes are clearly seen in this photograph. Other features are described in the text

high-resolution 13 cm tubes running at 50 kV EHT. It is intended for use with a concave screen having a light-gain of ×5. Dynamic convergence is handled by a microprocessor-based control system feeding separate correction coils on each tube.

High-definition TV summary

As our brief account has shown, the generation, distribution and display of high-definition TV pushes every aspect of TV design way beyond the limits imposed by the present TV system, in which the capabilities of all the links in the chain are fairly evenly matched. To bring them all up to standard for higher-resolution television pictures will cost a great deal of money, and require a great deal more research. Although all the technology for advanced TV already exists, and very impressive results have been demonstrated, the implementation of *domestic* systems (especially broadcast-derived ones) awaits many developments, mostly in the regions of squeezing more information into existing channel-widths; and the economics of manufacturing practical fine-detail display devices.

The bandwidth problem is one that will not be solved by money alone! One area which is being intensively researched is the idea of discarding redundant picture information, i.e. the stationary parts of the picture which are duplicated over a series of frames. Techniques have already emerged in which stationary and moving parts of the image are treated separately for noise and cross-colour reduction purposes, and in the long term a 'store and update' transmission scheme may offer the solution to the very greedy bandwidth demands of high-definition and Hi-fi TV systems.

3D TV

Perhaps the ultimate step in television technology is the presentation of a three-dimensional display with depth and perspective as well as full colour and high definition. One day this may be realised by means of *holographic* techniques in which interference patterns from laser light sources are able to create an apparent three-dimensional image without the need for the viewer to wear special spectacles or other optical aids. 3D systems using more conventional display sources have been demonstrated, however, and have even found their way onto domestic TV screens in the form of demonstration transmissions, for which special bi-colour viewing glasses are required.

One system uses two conventional TV sets, fed with video signals from two cameras viewing the same scene from different angles. The sets are mounted in 'L' formation with the screen planes at right-angles; each screen projects an image onto a sheet of reflective plate glass set at an angle of 45° between the two screens. Viewed through special spectacles the effect is very striking, with the picture appearing to project forwards into the space in front of the display surface.

An alternative approach, based on the stereoscopic principles known since Victorian times and exploited in some cinema films from time to time, involves a colour-separation process in which red images are separated from green (or cyan) and presented side-by-side on the display screen to render a picture which looks like a badly-converged colour display. The horizontal separation of the two coloured images determines the apparent depth and spatial position of the object thus outlined, and will vary to describe 'Z axis' (i.e. on a line between the viewer and the display surface) movement. This effect can be exploited in a basically black-and-white or colour picture.

A means of artificially inserting a 3D effect into ordinary colour broadcast (or videotape) pictures was introduced in 1983 by European setmakers Saba and Nordmende. When invoked, the 3D facility horizontally separates the red and green images by means of an electrical time delay, and a quasi-3D effect is seen when viewing the result through the red/green spectacles supplied. There are several alternative approaches to 3D TV short of the full holographic treatment, and if and when a common and compatible system emerges it may well involve the use of polarised light.

9

MAC encoding and satellite TV

With the coming of a new broadcast media such as cable and particularly satellite, whose bandwidth is less constricted than terrestrial broadcasts in the VHF and UHF spectra, much development work has been taking place on new encoding systems to provide displays free of the impairments described at the end of Chapter 4. Some offer the incidental advantages of potentially better definition and elimination of flicker and *aliasing* effects in the luminance picture.

A relatively simple one, proposed by the BBC was the *Extended-PAL* system, in which the inability of a conventional TV receiver to fully display the fine luminance detail (in the presence of colour) is acknowledged by transmitting a conventional PAL signal, but with the bandwidth of the luminance component restricted to about 3.5 MHz. This gives rise to little deterioration of the picture on a conventional set using PAL techniques and a shadowmask tube, where the presence of a luminance *notch-filter* and the shadow-mask (which acts to filter out fine detail, especially in the smaller screen sizes) virtually delete the finer detail anyway. Chrominance is transmitted in the conventional way, based on the subcarrier of 4.43 MHz, so that this scheme is a compatible one (see Figure 9.1), but without significant spectrum-sharing by the Y and C signals.

The high frequencies of the luminance signal, corresponding to fine detail, are sent separately, spaced from their rightful position by exactly the colour subcarrier frequency. Thus a 5 MHz luminance component will appear at $5 + 4.43 = 9.43$ MHz. In suitably-equipped receivers the re-generated subcarrier can be used to drive a second luminance detector whose output, after filtering, can be added to the basic 'restricted luminance' signal to reinsert fine detail and achieve full definition pictures in colour without spurious effects. The gap in the extended-PAL spectrum between about 6 and 8 MHz is free for up to six digitally-modulated sound signals riding in a subcarrier (Figure 9.1 again).

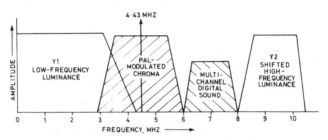

Figure 9.1. The spectrum of an E-PAL system. High-frequency luminance components (section Y2) have been shifted up in frequency to largely eliminate Y–C band-sharing

This solution offers the advantages of compatibility and a low-cost receiving and decoding system for the enhanced colour pictures. It is, however, totally tied to the PAL system, and would have committed us to PAL and chroma subcarrier techniques for the forseeable future.

Except in the sound channel, extended-PAL and similar systems take little account of the great advances which have been made in television technology in the decades since the subcarrier system was introduced into colour broadcasting, particularly in the area of colour picture encoding. With the economies of scale afforded by mass-production, digital and 'hybrid' ICs for use in television signal-processing can be made viable for domestic use in decoding systems which are more complex, but potentially capable of better performance.

166

Time-division multiplex

Where several signals need to be carried over a single transmission system, various possibilities are open to us. We have already explored the interleaving mode, which is only viable where the separate signals for dispatch have a common 'base frequency'. An alternative system is frequency-division multiplex, where the various signal channels are carried on different basic or carrier frequencies. Probably the simplest everyday example of this is the conventional TV aerial downlead which contains four TV programmes and accompanying sound tracks, all carried on different frequencies (channels) and easily sorted out by tuned acceptor circuits in the TV's front end.

A third alternative is the concept of *time-division multiplex*. Here the various streams of information are sent sequentially, each in a well-defined 'time slot' in the transmission. If we are to have a continuous signal stream available at the receiving end, this implies some form of storage system there to sustain the information flow during the periods when other information is present in the channel. A further requirement is that a synchronising signal be present at regular intervals in the composite signal to time the 'strobe' style detection and separation process.

Time-compression of analogue signals

Suppose we wish to send a five-minute spoken message over the telephone, perhaps to a far-distant country for which telephone time-charges are high. We may choose to compress our message into a time slot of 2½ minutes by means of recording it on tape and replaying into the 'phone at double speed. If our correspondent (suitably forewarned!) were to hook the telephone to a tape recorder running at twice normal speed he could capture the message for subsequent replay at normal speed. In this way the *real-time* transmission phase is reduced by half, though at the expense of bandwidth. Inevitably the twice-speed message contains twice the frequency spectrum of the uncompressed message, and the media (in our case the telephone link) must be able to carry this.

Time-division multiplex and time compression are two ingredients of the MAC transmission system. They are also used in the *digital* TV systems currently appearing in home TV and video equipment.

The digital video signal

The concept of quantising analogue waveforms was briefly described in Chapter 1, and following on from that we must now look in more detail at the way in which a TV picture signal is encoded in binary form. Figure 9.2 represents the video waveform over a period of $1\,\mu s$ which, on a typical domestic TV with a 56 cm diagonal screen, would occupy about 9 mm of *one* scanning line. The picture pattern on this section of the scanning line can be seen to start at mid-grey,

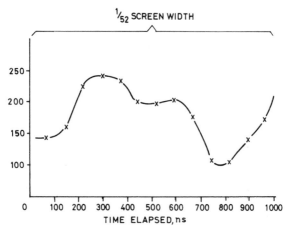

Figure 9.2. Quantising a video signal: the waveform represents a small portion of the video signal of one TV line. Quantising levels are represented by the crosses

brighten up, then pause at 'bright grey' before dropping to a lower level. At the end of our $1\,\mu s$ period the brightness is increasing again towards peak white. On the left-hand scale of the diagram is shown the quantising levels between 0 and 256, while along the bottom scale time intervals are shown up to $1\,\mu s$. At regular intervals of about 74 ns (ns, nanosecond, one thousand-millionth of a second) the video waveform is sampled to establish its level at that instant. In Figure 9.2, then, our samples at 74 ns intervals are 143, 160, 223, 241, 234, 200, 196, 203, 175, 105, 103, 140 and 170, making 13 samples. These numbers are produced in a fast A–D (analogue-to-digital) converter, not in the form of our decimal numbers, but in binary code; an eight-bit (bit, binary digit) word is used, with the necessary range of decimal 0–255. Binary coding is the simplest way of representing decimal numbers in 'pulse' form; reading from the *right* the 1s represent the numbers of 1s, 2s, 4s, 8s, etc. Our first three samples, then, are 10001111, 10100000 and 11011111, and so on throughout our $1\,\mu s$ 'window'. The digits come out in serial form, necessarily fast! All eight digits must be produced in the 74 ns period, so that for real-time digitisation we must expect around 100 bits per microsecond, representing a bit-rate of 100 Mbits/second, though the signal is now in binary form, and has only two states, 1 and 0. Various techniques can be used to reduce this bit-rate by about 30 per cent without impairment.

There are two approaches to quantising video signals. Either the complete video signal (encoded, with subcarrier, to PAL, NTSC or SECAM standard) can be quantised at a high sampling rate to render a composite digital signal, or (better) each of the Y, V and U signals can be separately quantised to give three digital pulse streams for combination before transmission. In this system, called *component coding*, the sampling rate for each of the colour-difference signals need only be about half that for the luminance signal due to their lesser bandwidth, and correspondingly lower *Nyquist* frequency. Because colour-difference signals, as we have seen, can have positive or negative values, zero is taken as level 128 from which the colour-difference signal can go down towards zero for negative signals and up towards 255 for positive signals. The total of 256 levels available imply an eight-bit word as before. The advantage of component coding is the compatibility of the system, in which each component, Y, V and U is effectively kept separate throughout the transmission and reception chain. Upon receipt, they can be fed separately to the RGB matrix and display channels of a colour receiver, or encoded in whatever form is required for the country or area intended.

Back, now, to our quantised video signal of Figure 9.2. At the end of the chain the digital pulse stream is presented to a D/A (digital-to-analogue) converter, whose output can describe 256 different levels, corresponding to the numbers conveyed in the bit-stream. This is the recreated facsimile of the original analogue signal, and at the sampling rate employed, it is a very good analogy indeed, as shown in

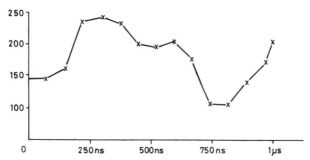

Figure 9.3. The video signal of Figure 9.2 reproduced after passing through a digital coder. It is a very good facsimile of the original signal

Figure 9.3. Over our 9 mm section of a single scanning line it is subjectively indistinguishable from the original signal, except for the complete absence of noise and other spurious effects! It is not hard to see that noise (and virtually all the noise – or snow – we see on our TV screens is picked up along the way) would not be reproduced on the digitally-transmitted picture; provided the 1s and 0s of the pulse

170

stream were distinguishable from each other in the receiver's decoder, no noise at all would be visible in the displayed picture. 6-bit quantisation is used in some domestic equipment.

Multiplexed Analogue Components (MAC)

In the outline of TV signal digitisation given above, there are about 860 samples per television line, each sample consisting of an eight-bit word. If we have the means of storing the 860 analogue samples we can 'hold' one TV line at a time in a *memory* consisting of a large number of CCD *(charge-coupled)* elements within an IC. We have to write in the samples at television scanning rate, but we can read them out at any rate we choose; see Figure 9.4. If we read them out at a faster rate we shall

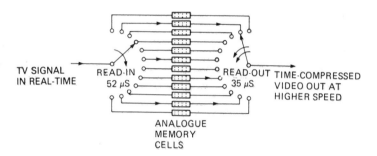

Figure 9.4. Reading in and out of a TV line store at different rates. Each memory cell forms a short-term store for a single pixel, and for a single TV line around 600–800 such stores would be required

effectively have compressed the signal in time at the expense of a faster signal-rate, rather like our telephone and tape-recorder trick described earlier. The active picture period of a 625-line transmission is 52 μs (as shown in Chapter 1) so we shall need this period to write one TV line of luminance into the CCD memory. Let's suppose we read it out in 35 μs, representing a time-compression of about 30 per cent. This will increase the data-rate proportionally, but leave a gap in the time-domain (assuming the standard TV-line 'window' of 64 µs) of 29 µs, in which we can do some time-multiplex magic to accommodate chroma and sound signals. Now for the chroma. Here we have about 18 µs available per line for transmission, and again we write the TV chrominance signal (say V, representing R − Y) into a one-line store. The read-rate here is faster than for luminance to squeeze 52 µs-worth of V information into a real-time slot of 18 µs, time-compressing the V-signal by about 65 per cent, again with a proportional increase in data-rate.

The now time-compressed luminance and chroma signals are assembled in the form shown in Figure 9.5 which clearly shows the time-division-multiplex composition of the MAC signal. As only one colour-difference signal can be accommodated per

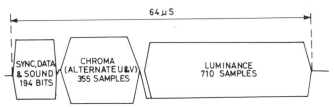

Figure 9.5. The signal format for MAC. Full details are given in the text

data-line, V and U signals are sent alternately on a sequential basis, rather like the SECAM system of colour encoding, where V and U signals are dispatched in the same way, though using a totally different modulation system, and operating in real time alongside the Y signal. Both SECAM and MAC require a one-line delay at the receiving end, not for cancellation-matrix purposes like a PAL decoder, but rather to yield a stored chroma line while its fellow is being received.

MAC sound and sync

Our 64 µs data line is now occupied by about 53 µs of picture signal information in 'compressed' form, first 18 µs of V or U, then

172

35 μs of Y. This leaves about 10 μs for sound and synchronisation signals. While this may not sound very much, the tremendously-high bit-rate inherent in the system affords almost 200 bits in this period, which, remember, need only sustain the sound for a 64 μs period. Thus we have room in the *line sync word* for an initial eight bits (to provide system synchronisation for sound and vision demodulation), then sound and data (teletext) signals, all in digital form. The MAC-C system has potential for eight simultaneous data channels which may be used to carry either sound tracks or teletext pages.

MAC decoding

It is not possible to go into the MAC decoding process in much detail in this book, but as with any form of decoding, the process must be the 'reciprocal' of that at the sending end. Thus we write the luminance and chrominance signals into CCD one-line stores at the fast transmission rate and simultaneously read out of them at normal 625/50 television rate to render continuous luminance signal and line-sequential V and U signals. By means of a 64 μs delay-line store and an electronic double-pole switch a constant flow of colour-difference signals is secured, each of R − Y and B − Y being updated at 128 μs (two-line period) intervals.

MAC interfacing

At the output of the MAC decoder we have sound and vision signals in baseband form. The latter may be RGB or Y, V, U signals, and ideally they should be fed direct to the picture-tube drive circuitry of the TV set (or monitor) in use to realise the full advantage of this mode of encoding. Initially the MAC decoder will form an add-on unit to the TV set, as an integral part of the down-converter for satellite reception, or the terminal unit of a wideband cable system. Where the TV receiver has sockets only for UHF or baseband

173

CVBS inputs it will be necessary to recode the chroma in PAL (or SECAM where appropriate in Europe) to enable the set to display a picture, which will necessarily then suffer all the shortcomings of cross-colour etc. related above! For older-type TV sets with just a UHF input (these are often 'live chassis' designs) a suitable RF modulator as used in VTRs, TV games etc., will be required as an output device in the conversion equipment, whereupon a spare channel in the receiver will be reserved for cable and satellite transmissions, and tuned to the modulator's output. More details will be given later in this chapter.

Subcarrier versus MAC signals

The three established encoding systems, NTSC, PAL and SECAM, were designed with the characteristics and particularly the noise performance of terrestrial VHF and UHF AM broadcast systems in mind, and with the need for compatibility with monochrome equipments at the top of the priority list. Many of the constraints which existed then have disappeared, and the main stumbling block – that of channel bandwidth – has considerably eased as carrier frequencies (and now bit-rates) have gone up and up. With the tremendous coverage area of satellite transmissions, a different form of compatibility now has to be considered, that of the different encoding systems – and languages – of the countries within the catchment area of a single transmitter in space, or indeed the vast distance over which TV signals can be propagated along fibre-optic lightguides. The language barrier is catered for by the availability of multiple sound channels, and the idea of component coding, in which each colour-signal is preserved intact throughout the transmission chain, gives great flexibility in the types and geographical spread of receivers in use, once the terminal equipment is freely and cheaply available.

Another advantage of MAC is the 'programmability' of its signal and synchronising systems, allowing for the first time

in broadcasting formats a degree of flexibility to accommodate future changes and improvements in television technology, without rendering obsolete a great percentage of the
total receiving installation.

Satellites

A relative newcomer to the broadcasting field is the practical
realisation of space transmitters. Before we attempt to
describe them in any detail, we need to understand a little of
how a satellite is positioned and controlled so as to appear
stationary to an earth-mounted receiving installation.

As we are well aware, the force of gravity is a dominant one
on earth, tending to pull loose objects towards the centre of
the earth. There is another effect called centrifugal force
which operates on spinning objects, this time pulling
outwards, away from the centre of the spin-arc, as
demonstrated by the domestic spin-drier, certain fairground
rides and a conker on a string. The strength of centrifugal
force depends entirely on the orbital speed of the object in
question. For any given orbital speed, then, there must be a
point in space where centrifugal force acting on an orbiting
satellite just cancels the gravitational pull of the earth, so that
these balanced forces hold the satellite at a fixed distance
above the earth's surface without any need for motive power
from the satellite itself. If we require the satellite to appear to
stand still relative to Earth, it will need to orbit at exactly the
same speed as Earth itself – once every twenty-four hours.
With the orbital speed thus rigidly fixed, the distance from
Earth for the satellite comes out at 35 800 km (22 200 miles) to
satisfy the gravity versus centrifugal force equation. So we
have two fixed parameters for our *geostationary* satellite,
orbital speed (which comes out at 11 069 km/hour for one
revolution per day) and the altitude of 35 800 km. It must
necessarily be over the equator to prevent an apparent daily
North-South oscillation. When all these parameters are
satisfied, the satellite will effectively hang stationary in the
sky. From time to time drift will inevitably take place of its

position and attitude, and this is corrected by small on-board jet motors which operate on a *servo* principle to stabilise and maintain the satellite's attitude to earth.

Thus we can set up a transmitter in the sky, but at first sight its position above the equator is bad news for the UK, situated as it is some way into the Northern Hemisphere. Plainly, a radio beam from the satellite will come in to us at an oblique angle rather than directly downwards; in the UK the

Figure 9.6. Satellite footprint: the primary and secondary service areas of the UK DBS satellite. The cross in the Irish sea represents the beam centre

elevation angle from the horizon varies from 17° in the Shetlands to almost 28° in West Cornwall. Except in built-up areas, a clear line-of-sight to the sky can usually be secured at this angle from ground level, so that receiving aerials can be mounted at ground level in many domestic situations, perhaps in a back or side garden; only in the shadow of buildings or other obstructions will it be necessary to mount the receiving dish on a wall, roof or tower. In urban areas where these problems exist they are more likely to be solved by the use of a 'community dish' with local cable distribution to individual homes. For those who wish to ascertain their own reception chances, the satellite position corresponds to that of the sun at about 3 pm BST (1400 GMT) on an early March or mid-September day.

The slant 'radio-illumination' effect stemming from the satellite's position over the equator at 31° west of Greenwich (assigned to the UK by world agreement) means that a narrow transmission beamwidth can cover the entire UK; for a *footprint* like that shown in Figure 9.6 the beam need only be 1.8 degrees in the 'vertical' plane and 0.7 degrees across.

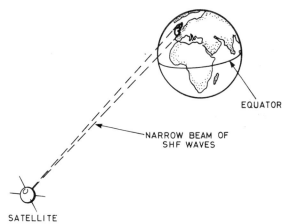

EQUATOR

NARROW BEAM OF SHF WAVES

SATELLITE

Figure 9.7. The relationship between the Earth, equator and a geostationary DBS satellite

Because all the transmission energy can be concentrated in this very narrow beam by highly directional (almost like a gun!) transmitting aerials at the satellite, the use of radio energy is very efficient indeed, though the narrow beam calls for very great accuracy in the alignment of the satellite in space, and the receiving dish in the garden! They have to be held within ±0.1° and ±0.5° of nominal pointing position respectively.

To give an impression of the radio-torchbeam-from-space effect, and the reason for the elliptic/potato shape of the footprint, Figure 9.7 represents the relative positions of the satellite, the equator, and the UK.

DBS bands and channels

Direct Broadcasting by Satellite (DBS) takes place in the SHF range on band VI, whose coverage in 11.727 to 12.476 GHz. There are five channels assigned to the UK, as follows: Channel 4, 11.78502 GHz; Channel 8, 11.86174 GHz; Channel 12, 11.93846 GHz; Channel 16, 12.01518 GHz; Channel 20, 12.09190 GHz. Three of these channels will be taken up by BSB (British Satellite Broadcasting) with technical help from the IBA and some programme input by ITN. Each channel is 27 MHz wide (compare with the 8 MHz bandwidth of terrestrial broadcast systems), giving plenty of elbow-room for the sidebands of the FM-modulated TV signal described earlier in this chapter.

The channel spacing in band VI is 19.18 MHz, so at first sight the allocated bandwidth is going to greatly overlap adjacent channels; so it does, but care has been taken to prevent interference by widely separating the recipient countries of adjacent channels. Thus, for example, the adjacent channels to our UK transmission have been assigned to Luxembourg – far enough away (especially when the *capture effect* of FM signals is taken into account) to prevent any interaction.

As with terrestrial broadcasts, the transmitted carrier wave from the satellite can be polarised in any required direction to minimise interference between adjacent and overlapping channels. By making the polarising angle continuously rotate through 360° we set up what is called circular polarisation, which is currently used in FM radio broadcasting to facilitate better reception on vertical car aerials. This is analogous to a spiral in effect, and shared channels are given opposite directions of circular polarisation, so that UK satellite broadcasts have right-hand circular polarisation, whereas

DBS signals for Andora and Austria, which share the UK channels, are given LH circular polarisation. Luxembourg has been assigned channels adjacent to those for the UK, and has RH circular polarisation, but rejection is assured by the different orbital positions of the satellites involved. Vision modulation by the C-MAC signal is FM without, of course, any additional sound carrier or subcarrier for the sound – it will be remembered that multiple sound channels are carried in digital form in the 'synchronisation word' of the C-MAC format.

Satellite power

In Chapter 5 we discussed the effect of a directional radiating aerial on the apparent power of the transmitter, and saw that ERP was the product of true RF power multiplied by the 'gain' of the transmitting aerial. The very narrow beam from a satellite (in the UK case, $1.8° \times 0.7°$) concentrates the power beautifully, and from an RF output power of about 150 W we can realise an ERP of 3000 kW! The only type of radiating aerial capable of launching such a narrow RF beam is a parabolic dish, and this is the basis of the satellite's transmitting element. It may be between 1 and 3 m in diameter, depending on the beam angle required. The true RF power of the satellite contrasts remarkably with that required for terrestrial broadcasts – on the one hand 150 W for complete coverage of the UK, on the other many many megawatts, indifferent coverage of some areas and over 650 individual transmitters to provide and maintain, in each case for one TV channel.

The power to run the transmitter and the peripheral equipment in the satellite itself comes from banks of solar cells drawing energy from the sun. These are mounted on 'wings' springing from the satellite body, and are servo-controlled to keep them at right-angles to the sun's rays. This accounts for the characteristic dragonfly shape of a broadcast satellite (see Figure 9.8). Inevitably the satellite will from time to time encounter a total eclipse when the earth blocks its view of the sun, and for the UK satellite this will occur around the spring and autumn equinoxes when for a short period each night the solar cells will be in shadow and incapable of powering the transmitter. Fortunately this will happen well after midnight, and for a maximum duration of 72 minutes and so can be accommodated within the programme schedules. This is fortunate indeed, preventing as it does the necessity for storage cells on board the satellite, with their weight and efficiency problems.

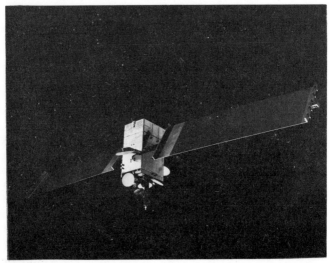

Figure 9.8. The DBS satellite: a model of L-Sat (photograph courtesy BBC)

To realise an RF power of 150–200 W, the DC power provided by the solar cells needs to be about 350–400 W per channel which implies around 1800 W required from the solar array to operate five channels. Much depends on the efficiency of the transmitter and the requirement of other, 'housekeeping' equipment on board. Some countries, due to their size and position relative to the equator, require greater beamwidth than others, and for larger beamwidths the RF power required increases rapidly. As an example, if the UK were directly below its satellite on the equator at 31°W (that is just off the Brazilian coast) we would need a fivefold increase in power to achieve the same field strength over the same area.

The satellite as an 'active reflector'

Of course, the programmes are not originated on board the spacecraft, so it must also be fitted with a receiving aerial and the necessary transposing equipment. Coming up to it from earth will be the programme material and signal routing instructions, along with engineering control signals to cater for 'housekeeping' and positional control. This uplink will

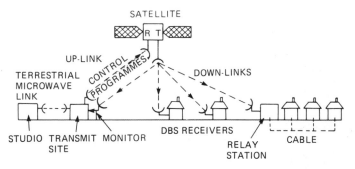

Figure 9.9. Arrangement of up- and down-links for the DBS satellite

also take place in the SHF band with a narrowbeam signal between parabolic dishes. Slotted into the satellite's output signals will be data streams containing information about the internal workings of the spacecraft and its motor, power and signal-processing sections for monitoring on Earth. The up- and down-links are shown in Figure 9.9.

The active life of the craft will come to an end when the fuel for the correction motors runs out, assuming that no catastrophe occurs to the other systems to disable the

181

satellite. As in all space-equipment design, crucial components are duplicated, and a great deal of potential redundancy is designed into every part of the vehicle. At the end, the satellite can be 'kicked off' into obscurity in deep space by a final burst of the positional motors at the very last, to make room for its successor!

Satellite signal reception

Leaving aside now the 'Star Wars' atmosphere of the orbiting satellite, it is time to return to the back garden to see what will be required in the way of receiving aerials for space transmissions at 12 GHz. While in theory a dipole with reflector and directors might work, the reality is that at the wavelength of a 12 GHz carrier (about 2.5 cm) the conventional aerial would be small enough to slip into a shirt pocket and would certainly not have enough 'meat' in it to extract a usable (or even detectable!) signal. We need to capture the beam over a relatively wide area and focus it on a dipole to achieve sufficient field strength to make a usable signal. For this a parabolic dish is called for, with some form of dipole at its focal point.

The larger the dish, the more RF energy it will intercept and concentrate on the pickup element; thus dish size has a counterpart in element-count in a conventional Yagi array. Again, the question of *phase* is relevant because all the reflected cycles of RF carrier must arrive in phase at the dipole in order to reinforce each other; see Figure 9.10. If any arrive out of phase a cancellation effect will take place with consequent loss of gain. This requires tremendous accuracy in the manufacture of the dish which must not deviate from a true parabolic shape, and must have a good surface finish. Any physical damage, warping or contamination of the dish surface will reduce the efficiency of the receiving aerial, which in good conditions should be about 55 per cent.

Detail design of SHF dishes depends on the individual manufacturer, but the DBS service is engineered to provide a carrier-to-noise ratio of better than 14 dB for 99 per cent of

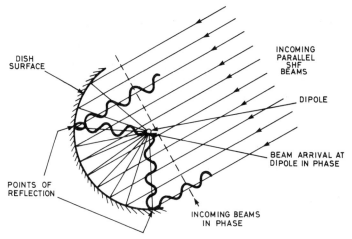

Figure 9.10. Parabolic receiving dish showing incoming parallel RF beam and the in-phase reflection to the pick-up device

the worst month with a dish of 90 cm diameter. A larger dish would offer improved gain, but be more difficult to align physically. Research is under way into the possibility of 'flat-panel' DBS receiving aerials which would be more aesthetically acceptable and easier to mount.

Head-end convertors

At the point of pickup, the signal is in 12 GHz FM form, and at that frequency the only possible transmission medium is a waveguide in the form of a large and rigid metal pipe. Any form of co-axial cable would introduce intolerable attenuation, so a frequency converter is mounted at the receiving dipole, in similar fashion to the masthead amplifier of some terrestrial signal-receiving installations. This converts the signal, by means of a heterodyne process, to a lower frequency around 1 GHz which is amenable to feeding indoors by means of very low-loss co-axial cable. At the receiver a second down-converter will be required for conventional TV sets; more of this later!

DBS service areas

Unlike conventional TV transmitters, the coverage of satellite broadcasting is not limited by constraints of line-of-sight, nor (except in a very local sense, as described earlier) by topographical features. For this reason the service area is much more homogeneous and predictable for DBS, falling off gradually from the centre of the footprint. Within the primary service area (see Figure 9.6) a 0.5 metre dish should suffice, and in the secondary service area (beyond the limits of the UK) a similar standard of reception can be achieved with a larger (higher-gain) receiving dish. By the same token, most parts of the UK are within at least the secondary service area of many European satellite services, so reception of these should be possible by means of a steerable (or second) dish suitably chosen and aligned.

Dish installation

Although dishes for use at 12 GHz are not so large as those required for UK reception of the earlier 4 GHz satellites such as the Russian Gorizont, a fair amount of weight and surface area are inherent in them. This means that their mounting must be very strong and stable, not only from a safety-hazard point of view in strong winds, but also because their very critical angular alignment can only be maintained if they are mounted on a very rigid foundation, ideally a concrete base in the ground.

Once a stable fix has been achieved the dish must be aligned onto the satellite's beam, which is rather like trying to shoot a bullet into the end of a metre-long 20 mm copper pipe at fifty paces, and get it to emerge at the exit without having touched the sides! The first step is to calculate the azimuth and elevation of the satellite at the receiving site, and this requires an exact knowledge of the position of the site, obtainable from an Ordnance Survey map, or a marine satellite navigator. From this the required angles can be calculated to several places of decimals and the dish aligned

Figure 9.11. A 0.9 m DBS receiving dish in a 'back-garden' setting (photograph by courtesy BBC)

with reference to a magnetic or gyro compass. What is for sure is that the traditional 'wave and watch' system used for conventional aerials working on terrestrial transmissions cannot be used – one group went on record as taking two weeks to find DTS-2 by trial and error! A typical 12 GHz receiving dish is illustrated in Figure 9.11.

Preamplifiers for SHF

We have seen that down-conversion is carried out at the dish by a superhet technique. Before mixing, however, a very low-noise preamplifier must be used to bring the carrier signal up to a suitable level. The noise generated in this first stage virtually determines the S/N ratio for the entire installation, and the performance of this amplifier is thus crucial. The FET (field-effect-transistor) is used here, and designs using complete FET SHF amplifiers integrated onto a substrate, IC style, have been developed to offer very low noise figures: less than 2 dB at a gain of 17 dB. The tuned circuits for selection and filtering are also etched onto the GaAs (Gallium Arsenide) substrate.

As SHF front-ends are developed with progressively lower inherent noise, advantage may be taken of this to reduce dish size (making alignment less critical) or alternatively to utilise the same size dish for reception of fringe (outer-footprint) reception of DBS transmissions intended for other areas of Europe. Some readers will be aware that for no perceptible noise on a conventional analogue-derived TV picture an S/N ratio of about 39 dB is required; with C-MAC digital encoding and FM modulation the carrier-to-noise ratio of 14 dB quoted earlier is quite adequate to provide a noise-free display picture.

The question of interference is a relevant one. Any interference to the up-link is very unlikely, and at the receiving site man-made interference is not significant at SHF, especially with the encoding and modulation systems used. Heavy rain at the receiving site will degrade the signal strength by about 4 dB, but in the primary service area this should not be detrimental to picture quality provided such factors as dish size and alignment and preamplifier gain are adequate.

Down-converter

Although a down-conversion process is necessarily carried out at the dish itself to reduce the carrier frequency to a more manageable one, the signal coming indoors is still incompatible with the TV, not so much in respect of carrier frequency (although currently a second down-converter to UHF is required) but in the modulation and encoding systems used. Several approaches are possible here, as follows.

(a) For use with a conventional broadcast TV set, the down-converter will contain a tuner, IF and detection circuits, then a MAC decoder. The RGB or YUV signals thus derived will then be encoded to PAL (for UK) or SECAM (for France etc.) and recombined with the baseband luminance signal before being applied to an *RF modulator* working in the UHF band. The modulator will also take the MAC-decoded sound signal, now at baseband, and compose from

186

them a UHF signal (corresponding to that of a terrestrial transmitter) for injection into the aerial socket at the back of the TV. Further, stereo outputs will be available for connection to a stereo hi-fi system.

Where the TV has single 'CVBS video' and 'audio' rear sockets, an alternative (b) would be as per (a) above, but without the RF modulator; this would offer a slight improvement in picture quality.

Where the TV set has RGB or YUV input sockets a third system (c) is possible. Here the MAC-decoded signals are fed directly to the picture-tube drive circuits, whereupon most of the advantages of the MAC system will become available and a noticeable improvement of colour picture quality will be seen. The fourth, and perhaps the best, alternative is the 'universal TV' concept, in which the television receiver itself would incorporate the functions of the indoor unit. It would contain a MAC decoder alongside the PAL one, and be capable of taking a signal direct from the dish-mounted frequency-converter. This sort of technology would be best implemented in a 'digital' TV set wherein most of the signal processing, and particularly the chroma decoding, is carried out on a PCM basis. A new generation of digital TV receiver would find a consummate match in a MAC-encoded DBS input signal. Stereo-equipped TV sets, in all the above cases, will be furnished with L and R audio signals. These alternatives are summed up in Figure 9.12.

Interfacing with other equipment

Now that we have seen how the DBS receiver interfaces with the TV itself, it's time to examine briefly its connections to other entertainment equipment in the home. First comes the VTR, which so far as the down-conversion gear is concerned, 'looks' like a TV set, with aerial and CVBS sockets. Either can be used, as described for the TV, but the resolution capabilities of home VTR machines in both the Y and C channels will ensure that a satellite transmission, when recorded and subsequently replayed, will appear virtually the

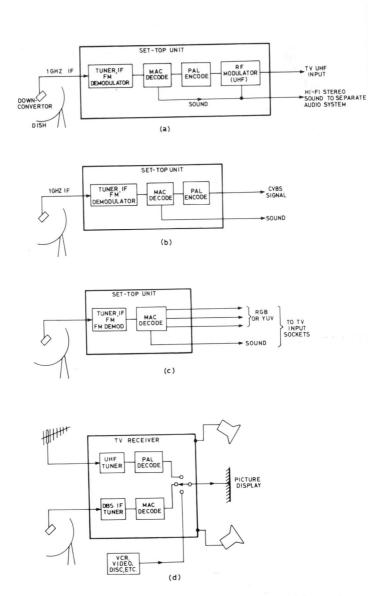

Figure 9.12. Alternative configurations for domestic DBS reception

same as a tape-recorded programme from any other source. S-VHS format gives good results, but for the older-established home video systems (offering the lowest-quality signal currently going), the 'off-air' DBS picture and sound quality will highlight the difference between live programme and replayed material. Because the audio signal will emerge from the down-converter in conventional baseband (L and R) form it will easily interface with the input of the stereo amplifier (or tuner-amp) of any audio hi-fi system. Once having been got into the system, as it were, conventional tape recording and reproduction of the DBS-derived sound signal is the same as for other signal sources such as disc or FM radio. For existing sound equipment the main problem is likely to be in finding spare input sockets and selector buttons if disc, tape and radio input facilities are all occupied! Regarding sound quality, there is every reason to believe that like the Compact Disc system, the main constraint on quality and S/N ratio in DBS-sourced programmes will not be the signal source at all, rather the amplifiers and particularly the loudspeakers in use, and the ambient noise level in the listening area – roll on the first live DBS coverage of the Last Night of the Proms!

Other satellites

A large number of non-DBS satellites are currently in orbit, offering a wide range of programmes, general and specialist. In the main they are supported by advertising revenue or by viewer-subscription, the latter often via cable-distribution companies. Most programmes are scrambled, requiring a licensed descrambler to make them intelligible. The exceptions carry heavy advertising and sponsorship, like Luxembourg's Astra satellite at 19.2°E. Astra programmes are MAC-encoded; many others use conventional PAL encoding with FM vision and sound modulation.

10

Cable television

Originally the philosophy of cable television was simple: to bring television to those who could not receive it by means of conventional RF aerials for reasons of local topography, the impracticability of installing an aerial, or aesthetics as judged by the local council or planning authority. Many CATV (Communal-Aerial Television) systems had their origins in the dark ages of 405-line television on VHF, with few and scattered transmitting sites. With the spread of the UHF terrestrial broadcasting network (transmitters are now being installed for communities of fewer than 300 souls) the necessity for CATV as an alternative to listening to the radio (!) has all but disappeared; blocks of flats, hotels and similar domiciles can be served by MATV (Master-Aerial TV) which is a small-scale cable system, working from a master aerial and distributing signals to tens, rather than thousands, of TV sets.

In some cases, out-of-area programmes were made available on a community cable network as a 'bonus' to subscribers, although the advent of Ch4 sometimes meant dropping this facility, as many cable systems (especially wired-pair HF and some co-axial VHF networks of long standing) had a maximum capability of four vision channels. Because of government policy, cable operators could, in general, only distribute the programmes of the national broadcasters, and this (in the UK at least) tended to limit the popularity of cable systems. Further problems for old cable networks were the propagation of teletext signals through the network, the difficulty of maintaining good bandwidth and delay characteristics for colour signals and the incompatibility of commercial home VTR machines with the special receivers (called *terminal units*) used with some cable systems.

In the 1980s interest in cable distribution systems was revived with the authorisation by the UK government of twelve pilot schemes for cable TV networks, each to serve about 100 000 households. Hence the great potential foreseen by the early pioneers, but not then permitted to develop, may be realised. The attraction of new cable franchises is that alongside 'off-air' material they will be permitted to broadcast exclusive programmes, not obtainable except over the cable, and thus create a demand from the viewing public and a financial incentive for the cable operators. It is envisaged that large-scale 'cabling' of the UK could follow, with its possibilities of Pay TV, and *interactive* services in which the viewer can communicate with a central information and data exchange via the cable, in the same way as Prestel viewers are able to send and receive videotext over the telephone network. The advent of satellite services opens further prospects for the cable system; those viewers who are unable to accommodate a receiving dish, or who wish to view programmes intended for other European countries (with or without English soundtrack) can be catered for, as the installation of receiving dishes for fringe satellite and DBS reception is most economically done on a 'community' basis.

Provided the demand (and hence finance) were there, the cable scheme would enable television to become as locally-based as the current BBC and ILR district radio services, particularly relevant in the provision of text and data transmissions of local interest only, and in the potential for local advertising.

Cable types

The early cable transmission system consisting of twisted pairs carrying HF vestigial sideband TV signals is now obsolescent, though still used for satellite programme relay.

The choice for multichannel TV carrying videotext (and possibly other data channels) lies between co-axial and glass-fibre optic cables, and each has various advantages and disadvantages. Co-axial cable is currently well established, and scores on the counts of easy interfacing with existing equipment and lower initial cost, at least on comparatively short runs. The optical-fibre technique requires more complex terminal equipment but has advantages in the areas of data-handling capabilities, immunity to electrical interference, security against 'tapping', and a small physical diameter, enabling a greater number of services to be laid in existing ducts, and a reduction in the cost of routing. Optical fibre has advantages over co-ax in terms of transmission efficiency, too; for a given data-rate the *repeaters* ('boosters' to overcome transmission losses) can be spaced at greater intervals than with the co-ax system. Regarding cost, glass is intrinsically much cheaper than copper and in volume production the glass-fibre technique may well show an overall cost advantage over copper cables.

It has been shown that for trunk lines optical transmission has much to recommend it, and British Telecom currently operates several glass-fibre optic links for transmission of television, audio, data and telephone traffic. Some can operate at very high data rates (140 Mbit/second) which confers the simultaneous ability to handle 1920 phone calls, or two broadcast-quality digital TV channels. It may be that the most economical way to implement a cable system will be to adopt a 'hybrid' solution, with optical-fibre trunk routes to local distribution points, whence co-axial cables will 'spur off' to individual dwellings grouped around the fibre cable head.

Transmission modes

In the same way as air or space can be used to carry virtually any radio frequency using any of the several modulation systems outlined in Chapter 1, so it is (within reason!) with co-ax and fibre-optic systems. Obviously the 'launching' and

192

'interception' methods differ, and for fibre the basic carrier is light, rather than an electrical wave. Thus we can use baseband, AM, FM, PM or PCM in cable systems, at such carrier frequencies as are appropriate to the signal, the distance between terminals, and the transmission medium. Typically a co-axial cable will carry analogue signals modulated by FM or AM onto carriers in the VHF and (for short runs) UHF range. Fibre-optic cables will work from analogue-baseband to the 140 Mbits/second PCM mode described above. It should be remembered that the light-carrier in a fibre system (usually infra-red rather than visible light) is itself an electromagnetic wave with a frequency of the order of 3×10^8 MHz or 300 THz (see Chapter 2) so that the upper limit on the rate of data throughput in an optical fibre, is perhaps, limited by technology rather than physics! The main restriction on bandwidth in glass fibre links using modulation of a sub-carrier (sub-, that is, to the frequency of the light wave itself) is the effect of *fibre-dispersion*, which tends to slightly 'blur' in time the sharpness of received pulses, giving an *integration* effect to their shape.

The network

There are two basic methods of cable distribution. The more traditional is the tree-and-branch system (see Figure 10.1a) in which all available programmes are continuously sent over the network in separate channels, with user selection by means of some form of switch at the receiving point. This was the *modus operandi* of the original radio and TV cable system, in which each household may be regarded as being on the end of a 'twig'.

The alternative and better system is termed a 'switched-star network', shown in Figure 10.1b. Here the available programmes are piped to a 'community-central' point anologous to a telephone exchange. The subscriber can communicate with the 'exchange' and request the desired programme(s) to be switched to his line for viewing, listening or recording. The advantage of the switched-star network is

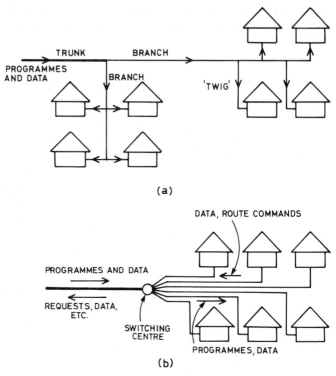

Figure 10.1. Two methods of signal distribution: (a) the 'passive' trunk-and-branch system; (b) the 'interactive' switched-star network. These are discussed in the text

its *interactive* capability, whereby the subscriber can 'talk-back' to a local or central exchange; this opens the possibility of the 'single-fibre' household in which all communications services (radio, TV, telephone, banking, public-utility meter reading, text and data etc.) come into the dwelling via a single link which can, by means of recording devices at either end, be utilised during off-peak and night hours. The switched-star configuration lends itself well to Pay-TV (either pay-by-channel or pay-by-programme) because security is more easily arranged. It is simpler to deny a programme to a

194

non-subscriber by a central 'turn-key' process than by expensive signal *scrambling* and 'authorised unscrambling' systems.

Terminal equipment and propagation modes in glass-fibre cables

Since the basic transmission 'vehicle' in fibre-glass cables is light energy we must now see how the signal is launched into the cable and intercepted at the receiving end. Depending on the distance to be covered, the sending device may be an LED or low-power semiconductor laser operating on a wavelength (infra-red) around 850 nm. The radiant energy in the sending device is surprisingly small, typically 200–300 μW for an LED and 1–3 mW for a laser. The fibre termination is an integral part of the light-source encapsulation for maximum coupling efficiency, permitting virtually all the light to be concentrated in the cable.

The receiving device is a light-sensitive diode, again intimately coupled to the fibre end. For low noise and highest possible sensitivity this pick-up device is 'tuned' to the light wavelength in much the same way as a radio set is tuned to an RF transmission. For short-haul reception a silicon P-I-N photodiode is generally used; with long-distance fibre cables greater sensitivity can be obtained by using an *avalanche* photodiode which combines the property of light detection with an internal amplification process.

The glass-fibre core is of very small diameter, typically 50–100 μm, surrounded by an intimately-bonded cladding layer of about 20 μm thickness. Further layers give strength and environmental protection, the outer jacket consisting of a tough water-proof polyurethane cover. The transmission of light along a glass fibre depends on the phenomenon of *total internal reflection* in which the light, when it encounters the inner surface of the fibre wall, is 'bounced' by the mirror-like wall surface back into the fibre. Light can enter the fibre end at any angle and the bounce-path in transmission can thus take several forms, as shown in Figure 10.2. Large bounce-angles give rise to a long path length and are called high-order transmission modes; at lesser angles the path length is shorter, described as low-order transmission mode. A beam which travels down the axis of the fibre takes the shortest possible path length in what is known as axial mode.

The nature of light propagation down a fibre depends on its diameter and on the difference in refractive index between the fibre core and its cladding material. Where a sudden change of refractive index is present at the fibre wall we have a *step-index* fibre in which several modes (high- and low-order) are taken by the light. An alternative form of construction is the *graded-index* fibre, where the interface

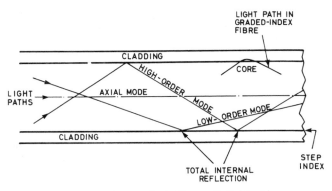

Figure 10.2. Propagation modes in fibre-optic lightguides

between 'core' and 'cladding' represents a more gradual change in refractive index. This has the effect of making the light rays turn less sharply when they encounter the fibre wall and thus reduces reflection loss, as shown at the top-right of Figure 10.2. Low-order modes predominate in such a graded-index fibre cable, and such high-order modes as are present travel faster along their longer path, reducing the fibre-dispersion effect described earlier. Graded-index fibres offer a low transmission loss and greater bandwidth than step-index types, at the expense of higher production costs and greater coupling losses at the junctions between the fibre and the sending and receiving devices.

196

Both step- and graded-index fibres operate in what may be called multimode with many possible light path angles within the fibre. If we can arrange a fibre to concentrate on the axial mode we shall significantly reduce the transmission loss. In this *monomode* system a high-grade glass core is used, with a small diameter in the region of $5\,\mu$m (about one-tenth the diameter of multimode cable cores). The light wavelength used here is longer, around $1.35\,\mu$m, and the much straighter light path gives very good transmission efficiency. Repeaters (regenerators) are therefore required at much longer intervals than in conventional fibre (and particularly co-axial) systems, and in a typical monomode transmission system repeaters can be as much as $30\,$km apart; this economy in equipment easily outweighs the cost disadvantage of the monomode fibre cable itself.

Repeater power

For repeaters generally, the operating power can be sent down the cable in co-axial systems, as described earlier for domestic masthead amplifiers. Fibre-optic cable plainly cannot carry DC, but conductive members can if required be incorporated in its protective sheath, or local power sources can be used, in view of the long intervals between repeaters made possible by fibre-optic technology. Many thousands of kilometres of fibre-optic cable are in use in the UK, primarily by British Telecom.

11

Development of video tape recording

The story of video tape recorders really began before the turn of the century with the experiments of Valdemar Poulson. By this time the relationship between electricity and magnetism was well understood, and the idea of impressing magnetic pulses on a moving magnetic medium was sufficiently advanced in 1900 to justify a US patent on Poulson's apparatus, the Telegraphone. The medium was magnetic wire rather than tape, and without any form of recording bias, and scant means of signal amplification, the reproduced sound signal was low, noisy, non-linear and lacking in frequency response. These problems of 'tape' and record/replay head performance are ones that have continually recurred throughout the history of sound and vision tape recording, as we shall see.

By the early 1930s, many advances had been made in the field. DC bias, or pre-magnetism of the recording wire had been tried with better results, then overtaken by the superior system of AC bias, as used today. The magnetic wire gave way to steel tape 6 mm wide travelling at 1.5 metres per second, and performance became comparable with the contemporary disc recording system. Not 'hi-fi' by any means, but certainly adequate! The BBC adopted and improved the Blattnerphone system and in 1932 broadcast a programme of the Economic Conference in Ottawa, for which *seven miles* of steel tape was used, edited by means of a hacksaw and soldering iron! This era also saw the first crude forerunner of a servo system in the Marconi–Stille machine of 1934. The earthed tape was arranged to contact insulated metal plates when it became slack, the plates being wired to thyratron control valves. Relays in the thyratron anode circuits modified drive motor currents to regulate tape speed.

A great impetus was given to the industry when it became possible to coat a flexible insulated base with a finely-divided magnetic substance. This was achieved in Germany by Dr. Pfleumer and developed by Wilhelm Gaus under the auspices of AEG. This activity culminated in the successful demonstration at the 1935 Berlin Radio Exhibition of the first commercial sound tape recorder, the AEG Magnetophone, using a cellulose acetate tape coated with carbonyl iron powder. Performance of these sound recording machines steadily improved during the '30s and '40s to the point where at the end of the 1940s much radio broadcast material was off-tape, and indistinguishable from live programmes.

Once audio magnetic recording had become established in the radio industry, attention was turned towards the possibility of recording television images on tape. The problems were formidable, mainly because of the relatively large bandwidth of a television signal. Plainly, it would be necessary to increase the tape speed and one approach, by Crosby Enterprises in 1951, took the form of a 250 cm/second machine designed to record a monochrome picture whose frequency spectrum was split into ten separately-recorded bands with additional sync and control tracks. The same principle was embodied in a fearsome machine designed by RCA, in which the tape travelled at 600 cm/second to record on three separate tracks simultaneous R, G and B information for colour TV. Other designs involving longitudinal recording called for tape speeds approaching 1000 cm/second, amongst which was the BBC's VERA (Video Electronic Recording Apparatus) of 1956. Such machines were wasteful of tape and thoroughly frightening to anyone who happened to be in the room in which they were operating!

Already the seeds had been sewn of a new system, one which was to hold the key to the modern system of TV tape recording. This was simply the idea of moving the record or replay heads rapidly over the surface of a slowly-moving tape to achieve the necessary high 'writing ' speed. Initially, the hardware consisted of a circular plate with three recording heads mounted near its edge at 120° intervals, their tips protruding from the flat surface of the faceplate. The 5 cm-wide tape was passed at 76 cm/second over the surface of the rotating plate, whose heads had an effective velocity of over 6000 cm/second, resulting in narrow arcuate tracks across the width of the tape. It was a step in the right direction, but results were poor for several reasons. The bandwidth of the TV signal being recorded was difficult to get on and off the tape due to noise and head-gap problems (we will meet these in detail in the next chapter), and the valve technology of the time did not really lend itself to such requirements as an ultra-wideband, high gain and stable amplifier.

Two more factors were required for success, and these were engineered by Dolby, Ginsburg and Anderson, of

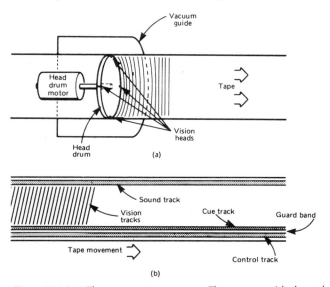

Figure 11.1.(a) The transverse-scan system. The vacuum guide draws the tape into a curved profile to match the contour of the head drum. (b) Resulting tracks on the tape

Ampex between 1952 and 1955. The problems associated with the tape-track configuration were solved by the use of a horizontal head-drum, containing four heads and rotating on a shaft mounted parallel to the direction of tape motion as shown in *Figure 11.1a*. The video heads lay down parallel tracks across the tape width, slightly slanted, (see *Figure 11.1b*) with each head writing about 16 television lines per pass. The tape-to-head speed (i.e. the *writing speed*) at an incredible – to us today – 40 metres per second was most adequate, with response beyond 15 MHz. The final hurdle was cleared with the introduction of an FM recording system; this involves frequency-modulating a constant-amplitude carrier with the picture signal before application to the recording head. Charles Ginsberg, one of the collaborators, once said that FM was proposed by the man who was assigned to design an AGC circuit for the AM system originally in use! The Ampex *Quadruplex* system was enthusiastically taken up by TV broadcasters, and rapidly became a world standard.

For less exacting requirements and limited funds, a simpler recording system was really needed, and this gave birth to the helical-scan system. The idea of a rotating video-head drum containing one or two heads, each laying down one complete television field per pass, was first mooted in 1953. The Japanese Toshiba company were foremost in the field in the early days, though several other companies were working on the idea. By 1961 a handful of manufacturers were demonstrating helical machines, all with the open-reel system, each totally incompatible with all the others, and none having a performance which could approach that of the quadruplex system. The market for these machines was intended to be in the industrial and educational spheres, with very little regard as yet to the domestic market. If there was one thing these professional users wanted above all else, it was standardisation of the helical format to ensure software compatibility. Does this, almost 30 years later, have a familiar ring about it?

By the early 1970s, helical-scan machines had, by and large, settled into three camps. The Sony U-matic standard was well established, with a very good performance for its era, using

12.7 mm tape in a large cassette for easy handling. Most Far-Eastern manufacturers had thrown their weight behind the EIAJ (Electronics Industries Association of Japan) system, using a cartridge containing a single spool of tape. The third contender was Philips. Their 'VCR' machine, illustrated in *Figure 11.2* was the first to be designed truly for the domestic as well as the educational and industrial markets. It took a relatively compact, fully-enclosed cassette and was styled for the home environment. Easy-to-operate controls, a simple timer facility and one-hour capability were offered, together with built-in TV tuner and RF modulator for use with a conventional TV set. This, the N1500, was a milestone in VTR development.

Figure 11.2. The Philips VCR format was first in the domestic field. A late version of the first Philips machine, type N1500

The N1500 did not sell in very great numbers, in spite of the intensive effort made by Philips' engineering and publicity departments. Its release in 1972 coincided with an explosion in colour TV sales, and the average householder blew his savings on a colour receiver rather than a VTR. The maximum one-hour playing time, insufficient for a football match or a feature film, also held back sales and the introduction by Philips in 1976 of a similar machine, but with two-hour capability (N1700, VCR-LP) was a step in the right direction. The Philips monopoly was broken, however, in 1978 when the rival Japanese systems VHS (Matsushita) and Betamax (Sony) appeared on the shelves of high street shops.

All three systems used cassettes, a thick double-stacked type for the Philips VCR and VCR-LP format, and smaller, lighter, co-planar cassettes for the VHS and Betamax systems. All the early machines of each format had piano-key controls similar to those of audio tape recorders. The rather awkward and expensive Philips VCR cassette, together with its relatively short playing time, spelt the early demise of this format, and by 1980 VHS ruled supreme in the UK, with Betamax in second place and rapidly gaining ground. Sales of domestic VTRs rapidly increased, and a wide range of software appeared alongside. The pace of machine development quickened, and in 1980 piano-keys had given way to light-touch sensors and remote control, sophisticated timers and programmers had appeared, and a form of freeze frame had been introduced. Another significant event of 1980 was the appearance of a new helical VTR format, the Philips/Grundig Video 2000 system, for use with the VCC (Video Compact Cassette). The Compact Cassette designation indicated the intention that it should become as popular as the universally-used audio compact cassette. In the event V2000 format was not successful, and production had ceased by early 1986. It used an advanced DTF (Dynamic Track Following) tracking system, very similar to that of the later and more successful Video 8 format, whose *ATF* feature will be described later.

Specialised ICs have been making steady inroads into domestic VTR design, LSI (Large Scale Integration) and microprocessor devices becoming commonplace in contemporary machines. These have made possible advanced remote control systems, comprehensive timers and programmers, trick-speed, still-frame and visual search features. They have also led a trend away from mechanical complexity in the tape deck and towards electronic control of direct-drive systems; this has considerably simplified the mechanics of the tape transport, threading and head drive systems, while retaining a relatively low electronic component count.

Advances have been made, too, in the field of portable VTR equipment. Purpose-designed battery operated machines are available in VHS, VHS-C and Video-8 format for outdoor location work. Those currently on offer incorporate the camera section and videorecorder in one unit – *camcorders*. In this realm there is strong competition between the VHS camp (consisting mainly of 'traditional' VCR manufacturers) and the Video 8 protagonists, led by Sony of Japan and having in its ranks many companies with backgrounds in the world of conventional and cinephotography. Details of VHS, VHS-C and Video-8 formats will unfold throughout the book, and the relative merits of the formats (including the high-definition VHS variant, *S-VHS*) will be discussed in detail in Chapter 21.

Magnetic tape basics and video signals

All magnetic materials, videotape coatings amongst them, may be regarded for practical purposes as consisting of an infinite number of tiny bar magnets, each with its own north and south poles. This is a simplification, but suits our purposes well. In the natural state these bar magnets are randomly aligned within the material so that their fields cancel one another out, and no external magnetic force is present. Thus the contents of a box of steel nails, for instance, will have no particular attraction for each other. To magnetise the material, but it a blank tape or a solenoid core, we have to apply an external magnetic force to align the internal magnets so that they sit parallel to one another, with all their N poles pointing in the same direction. When the external field is removed, most of the magnets remain in alignment and the material now exhibits magnetic properties of its own.

Remanent magnetism

If the relationship between externally applied force, or *flux* and retained flux in the material, were linear, the business of tape recording would be much simpler. Unfortunately this relationship, called the *transfer curve*, is very far from linear, as *Figure 12.1* shows. Here we have a graph with the magnetising force (H) plotted along the horizontal axis, and the stored flux density (B) on the vertical axis. Our starting point (with

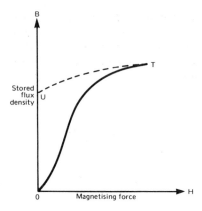

Figure 12.1. Initial magnetisation curve of a ferro-magnetic material

the new box of nails!) is in the centre, point O. Here, no external magnetising force is present, and the magnetic material's internal magnets are lying in random fashion – hence zero stored flux. Let's suppose we now apply a linearly-increasing external flux. The *flux-density* (strength of magnetism) in the material would increase in non-linear fashion as shown by the curve O-T. At point T the material has reached magnetic saturation and all its internal magnets are rigidly aligned with each other – no increase in applied magnetic force will have any effect. If we now remove the magnetising force, bringing the H coordinate back to zero, we see that a lot of magnetism is retained by the specimen, represented by point U. This is the *remanence* of the material, the 'stored charge' as it were. For recording tape, it needs to be as high as possible.

Hysteresis

Figure 12.2 is an expansion of *Figure 12.1* to take in all four quadrants. We left the material at point U with a stored flux. To remove the flux and demagnetise the material (we'll call it

tape from now on) it's necessary to apply a negative magnetising force, represented by O-V, whereupon stored flux B
returns to zero. Further negative applied force (V-W) pushes
the stored flux to saturation point in the opposite direction,
so that all the magnets in the tape are once again aligned, but

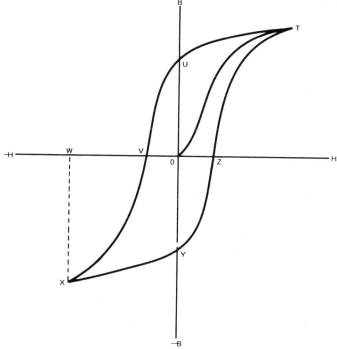

Figure 12.2. Hysteresis loop

all pointing the other way, represented by point X. Removal
of the applied force takes us back to the remanence point,
this time in the negative direction – point Y. To demagnetise
the tape a positive applied force O-Z is required, an increase
of which will again reverse the stored flux to reach saturation
once more at point T. The diagram is called a magnetising
hysteresis curve, and one cycle of an applied magnetising

waveform takes us right round it. It can now be seen how the 'degaussing' or *erasing* process works. Here we apply a large alternating magnetic field, sufficient to drive the material to saturation in both directions. The field is then allowed to decay linearly to zero, creating smaller and smaller hysteresis loops until they disappear into a dot at point O, and the material is fully demagnetised.

Transfer characteristic and bias

Figure 12.3a is based on the previous diagram, but shows the initial magnetising curve in solid line (TOX) with the hysteresis curve in dotted outline. If we apply a magnetising force, 1H, and then remove it, the remanent flux falls to a value 1'. A larger magnetising force 3H will, if applied and removed, leave a remanent force 3', and so on, up to saturation point 8H and remanence point 8'. The same applies in the negative direction, and plotting remanence points against applied force we get the curve shown in *Figure 12.3b*. This is the transfer characteristic. It is very non-linear at the middle and ends, but for a typical recording type, will contain reasonably linear sections on each flank, PP and QQ. It we can bias the recording head to operate on these linear parts of the transfer characteristic, the reproduced signal will be a good facsimile of that originally recorded. DC bias (*Figure 12.4*) puts the head into *one* of the linear sections, but the recorded signal will be noisy and inadequate; AC bias, shown at the top of *Figure 12.4*, allows the head to operate in two quadrants, with superior results. We shall see that the chrominance signal in a VTR is recorded along with a bias signal, which is in fact the FM carrier for the luminance signal.

Head-tape flux transfer

The recording head consists of a ferrite 'ring', with its continuity broken by a tiny gap. A coil is wound around the

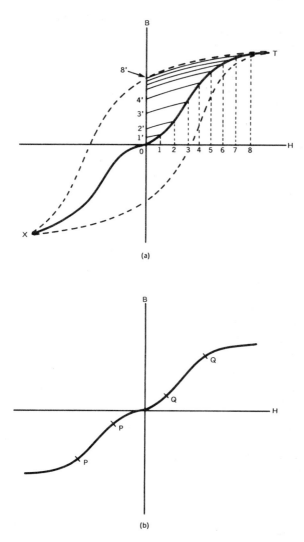

Figure 12.3. Deriving the transfer curve: at (a) is plotted the remanent flux for eight linear steps of applied magnetising force; (b) shows the resulting transfer curve

209

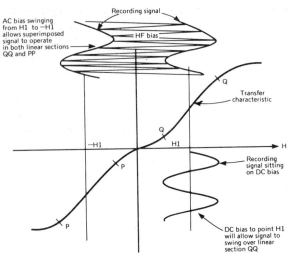

Figure 12.4. The effect of AC and DC bias

ring, and when energised it creates a magnetic field in the ring; this is developed across the head gap. As the tape passes the gap, the magnetic field embraces the oxide layer on the tape and aligns the 'internal magnets' in the tape according to the electrical signal passing through the head. Provided that some sort of bias is present (*Figure 12.4*) the relationship between *writing current* and flux imparted to the tape is linear, so that a magnetic facsimile of the electrical signal in the head is stored in the tape, as shown in *Figure 12.5a*. The tape passes the head at a fixed speed, so that low frequencies will give rise to long 'magnets' in the tape, and high frequencies short ones.

Head gap and writing speed

The linear relationship between head field strength and stored flux in the tape, described above, holds true when the wavelength to be recorded on tape is long compared to the

210

width of the head gap. However, when the wavelength of the signal on the tape becomes comparable with the head gap width, the flux imparted to the tape diminishes, reaching zero when the recorded wavelength (or rather the non-recorded wavelength!) is equal to the width of the head gap. This is illustrated in *Figure 12.5b*, where it can be seen that during the passage of a single point on the tape across the head gap, the applied flux has passed through one complete cycle, resulting in cancellation of the stored flux in the tape.

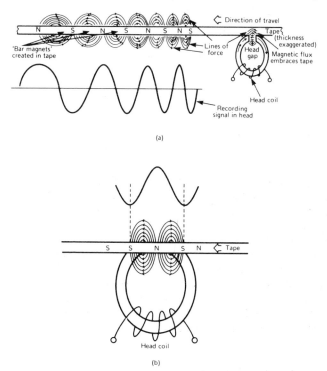

Figure 12.5. Storing flux on the tape. At (a) the flux appearing across the head gap is penetrating the oxide surface to leave magnetic patterns stored on the tape . (b) shows the effect when one complete cycle of the recorded waveform occupies the head gap - no signal transfer will take place

This point is known as the extinction frequency (Fex) and sets an upper limit to the usable frequency spectrum.

For video recording we need a large bandwidth, and a high Fex. This can be achieved by reducing the head gap width or alternatively increasing head-to-tape, or *writing* speed. There is a practical limit to how small a gap can be engineered into a tape head, and currently this is about 0.3 μm, less than half a micron (1 micron = 10^{-6} metre). For domestic applications, where full broadcast-bandwidth signals are not required, a writing speed of about 5 m/second is required with such a head-gap. How this is achieved will be explained in due course. Other HF losses also occur during recording. The head is by definition inductive, so losses will increase with frequency. Eddy currents in the head will add to these losses, as will any shortcomings in tape-to-head contact. High frequencies give rise to very short 'magnets' in the tape itself,

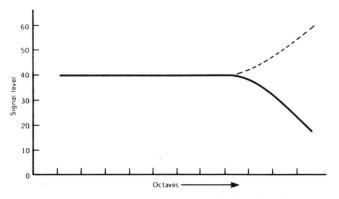

Figure 12.6. Losses in the recording process. The dotted line shows a compensating 'recording equalisation' curve

and it is the nature of these to tend to demagnetise themselves. For all these reasons, the flux imparted to the tape tends to fall off at higher frequencies, as in the solid line of the graph shown as *Figure 12.6*. To counteract this, recording equalisation is applied by boosting the HF part of the signal spectrum in the recording amplifier, as per the dotted line.

This is called recording equalisation, and its aim is to make the frequency/amplitude characteristic of the signal stored on the tape as flat as possible.

Replay considerations

As with any magnetic transfer system, the output from the replay head is proportional to the *rate of change* of magnetic flux. Thus, assuming a constant flux density on the recorded tape, the replay head output will double for each doubling of frequency. A doubling of frequency is called an *octave* and a doubling of voltage represents an increase of 6 dB. This holds good until the extinction frequency is approached, when the head output rapidly falls towards zero. This is shown in *Figure 12.7*. The upper limit of the curve is governed by the

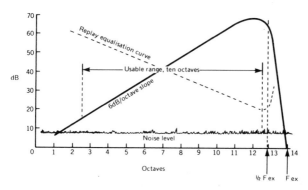

Figure 12.7. Playback curve for a tape with equal stored flux at all frequencies. For a 'level' output the gain of the replay amplifier must follow the dotted replay equalisation curve

level of the signal recorded on the tape, limited in turn by the tape's magnetic saturation point. At the low-frequency end, the replay head output is low due to the low rate of change of the off-tape flux. At some point, it will be lost in the 'noise' off-tape, and this will occur at about 60 dB down from peak level. Thus, even with playback equalisation (represented by

213

the dotted curve in *Figure 12.7*), the *dynamic range* of the system is confined to 60 dB or so. With the unalterable 6 dB/octave characteristic, we are limited, then, to a total recording range of ten octaves. This is inherent in the tape system and applies equally to audio and video signals. Ten octaves will afford an audio response from 20 Hz to 20 kHz, which is quite adequate. TV pictures, however, even substandard ones for domestic entertainment, demand an octave range approaching 18 and this is plainly not possible. No wonder they had so much trouble in the pioneering days!

Modulation system

To be able to record a video signal embracing 18 octaves or more it is necessary to modulate the signal on to a carrier and ensure that the octave range of the carrier is within the capabilities of the tape recording system. While the carrier could be AM (amplitude modulation), the FM (frequency modulation) system has been adopted because it confers other advantages, particularly in the realm of noise performance. An FM signal can be recorded at constant level regardless of the modulating signal amplitude, so that head losses and the effects of imperfect head-to-tape contact are less troublesome. To achieve a picture replay with no perceptible background noise (snow), the signal/noise ratio needs to be about 40 dB, and this can be achieved by an FM recording system in a domestic VTR. Professional and broadcast machines can do much better than this!

FM basics

An FM system, familiar to us in VHF sound broadcasts, starts with a CW (continuous wave) oscillator to generate the basic carrier frequency. The frequency of the oscillator is made to vary in sympathy with the modulating signal, audio for VHF sound transmitters, video for VTR recording systems and satellite broadcasts. For any FM system the *deviation* (the distance that the carrier frequency can be 'pulled' by the modulating signal) is specified. In VHF sound broadcasting it is ±75 kHz, giving a total frequency swing of the carrier of 150 kHz. In a VTR, carrier frequencies are specified for zero video signal amplitude – represented by the bottom of the sync pulse – and full video signal amplitude, i.e. peak white.

All modulation systems generate sidebands, and those for FM are more complex than occur in an AM system. In fact an FM system theoretically generates an infinite number of sidebands, each becoming less significant with increasing distance from the carrier frequency. *Figure 12.8* shows the

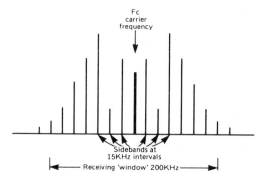

Figure 12.8. Sidebands of a VHF-FM sound transmitter

sidebands of a VHF-FM sound broadcast transmission. The modulating frequency is 15 kHz, and the sideband distribution is such that the first eight sidebands on either side of the carrier are significant in conveying the modulation information. Thus to adequately receive this double-sideband transmission we need a receiving bandwidth, or window, of 240 kHz or so – in practice 200 kHz is sufficient, and this is the allocated channel width.

Modulation index

The example given, 200 kHz bandwidth for transmission of a 15 kHz note, seems very wasteful of spectrum space, and certainly will not do for our tape system in which elbow-room is very limited! In the above example, the *modulation index*, given by the formula

$$\frac{\text{carrier deviation}}{\text{modulating frequency,}} \text{ is } \frac{75}{15} \text{ or 5.}$$

If we can reduce the modulation index, the significant sidebands draw closer to the carrier frequency, and at modulation indexes below 0.5, the energy in the first sideband above and below the carrier becomes great enough for them to convey all the necessary information in the same way as those of an AM signal. In video tape recording we go a step further and use only one sideband along with a part of the other, rather similar to the vestigial sideband scheme used with AM television broadcast systems.

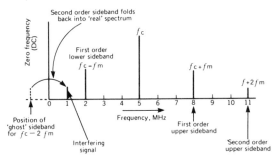

Figure 12.9. The effect of folded sidebands. Carrier frequency is 5 MHz, modulating frequency 3MHz

By using a low-modulation index, then, the sidebands of the FM signal can be accommodated on the tape. FM deviation has to be closely controlled and carrier frequencies carefully chosen to avoid trouble with the sidebands, which if they extend downwards from the carrier to a point beyond zero frequency will not disappear, but 'fold back' into the usable spectrum to interfere with their legitimate fellows, leading to beat effects and resultant picture interference. The effect of a folded sideband is shown in *Figure 12.9*.

FM video

Taking the VHS system as an example (all the formats use similar frequencies and parameters), the video signal is modulated on to the FM carrier according to *Figure 12.10*.

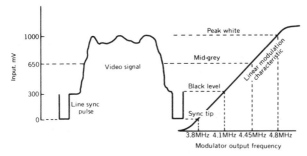

Figure 12.10. FM modulation characteristic for a VTR

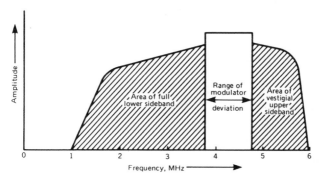

Figure 12.11. Spectrum of luminance video signal on tape

Sync-tip level gives rise to a frequency of 3.8 MHz, black-level 4.1 MHz, mid-grey 4.45 MHz and peak white 4.8 MHz. Thus deviation is limited to a total of 1 MHz, and with a restricted video bandwidth of about 3 MHz, modulation index is about 0.3. The resultant spectrum of the on-tape signal is shown in *Figure 12.11,* with a full lower sideband, vestigal upper sideband (limited by the system frequency response and the approach of

Fex) and a carefully arranged gap between 0 and 1MHz into which (as later chapters will show) is shoe-horned the chrominance, and where applicable ATF (Automatic Track-Finding) control signals. If we regard the FM luminance spectrum as extending from 1MHz to 6MHz, it is now occupying less than three octaves, well within the capabilities of the system. The S-VHS parameters will be discussed later.

Pre-emphasis

Noise is the enemy of all recording and communications systems, and in FM practice it is common to boost the HF components of the modulating signal prior to the modulation process. FM radio uses this technique, called pre-emphasis, and so do we in VTR FM modulation circuits. The effect of

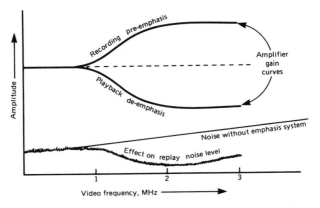

Figure 12.12. The effect of pre-emphasis and de-emphasis on playback signal-to-noise ratio

this after demodulation (in the post-detector circuit of a radio or the playback amplifier of a VTR) is to give a degree of HF lift to the baseband signal. In removing this with a filter, gain is effectively reduced at the HF end of the spectrum, with an accompanying useful reduction in noise level. The idea is

218

shown in *Figure 12.12*, which illustrates the filter characteristics in record and replay, along with the effect on the noise level.

Summary

The FM system is fundamental to the video recording process. Here is a summary of the main points, as applied to domestic VTRs. The video signal is bandwidth-limited in a low-pass vilter to restrict it to the band 0–3 MHz. This luminance signal (chroma components are removed by the filter) is pre-emphasised and applied to a voltage-controlled oscillator to provide an FM signal with a total swing of about 1 MHz, which is then limited and carefully controlled in amplitude to fall in the middle of the linear portions of the tape transfer characteristic. The FM oscillator frequency is chosen to minimise sideband interference and beat effects, and the FM signal is applied to the video heads after equalisation to compensate for HF losses. The signal is recorded on the tape at a writing speed of around 5 m/second, and the recorded tape carries a vestigial sideband signal ranging up to 6 MHz or so.

Video tape: tracks and transport

In Chapter 11 we saw that the necessary high head-scanning speed was first achieved by fast longitudinal recording techniques, then in the successful Quadruplex system by transverse scanning of the tape by a four-head drum with the video heads mounted at 90° intervals around its periphery. The transverse-scan method has a lot going for it! The very high writing speed confers great bandwidth, enabling full broadcast-specification pictures to be recorded and replayed. The relatively wide tracks are almost at right-angles to the tape direction, so that any jitter or flutter in the tape transport has little effect on the timing of the video signal, merely causing slight momentary tracking inaccuracies, easily catered for by the wide track and the FM modulation system in use.

The cost of transverse-scanning machines is high because of the need for a complex head-drum system and precision vacuum tape guides. To maintain the necessary intimate contact between video head and tape, all VTRs have their head tips protruding from the surface of the drum so that they penetrate the tape and create a local spot of 'stretch'. Hence the need for a precise vacuum guide at the writing/reading point in a transverse machine to maintain correct tape tension. These techniques are not amenable to domestic conditions or budgets, and the alternative and simpler helical system has undergone much development. Helical performance has reached a stage where it has challenged the transverse system in professional and broadcast fields, and the helical principle is embodied in all home VTR formats.

Principle of helical scan

In a helical scan system, several problems are solved in one go, but other shortcomings are introduced. The idea is to

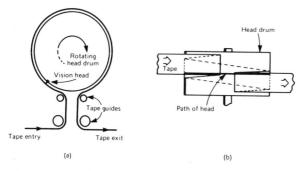

(a) (b)

Figure 13.1.(a) The Omega wrap. This diagram shows the arrangement for a single-head machine; the path of the head is the heavy line in (b) (b)

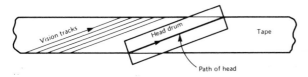

Figure 13.2. The diagram of *Figure 13.1* drawn from the point of view of the tape

wrap the tape around a spinning head drum, with entry and exit guides arranged so that the tape path around the head takes the form of a whole or part-helix. The principle is

221

shown in *Figure 13.1*, where a video head in the course of one revolution enters the tape on its lower edge, lays down a video track at a slant angle and leaves the upper edge of the tape ready to start again at the tape's lower edge on the next revolution. During the writing of this one track, the transport mechanism will have pulled the tape through the machine by one 'slant' video-track width, so that track no. 2 is laid alongside the first; each track is the width of the video head. *Figure 13.2* shows how this works and gives an elementary impression of the track formation.

Let's look at the strengths and weaknesses of the helical format. Tape tension and hence head-tip penetration is now governed by the tape transport system rather than a precision guide arrangement. One or two complete television fields can be laid down per revolution of the head drum, so that the problems of matching and equalising heads (to prevent picture segmentation) disappear. In a single-head machine such as we have described, no head switching during the active picture period is necessary. Two-head helical machines have a simple head-switch or none at all; what price do we pay for these advantages? The main penalty is timing jitter in the recorded an replayed video signals. Because the tracks are laid at a small angle to the tape direction (about 5° off horizontal for VHS) they may be regarded as virtually longitudinal, so that the effects of the inevitable transport flutter, variations in tape tension, bearing rumble etc will be to introduce minute timing fluctuationg into the replayed signal. This effect cannot be eliminated in any machine, and causes problems with colour recording and certain types of TV receiver, as will become clear in Chapters 15 and 20.

In practice, two video heads are used in the drum of a helical VTR and they can be seen in the photographs of Sony (Beta) and Panasonic (VHS) head drums (*Figure 13.3*). The two-head system means that the tape needs only to be wrapped around half the video head drum perimeter, with one head joining the tape and beginning its scan as the other leaves the tape after completing its stint. To give a degree of overlap between the duty-cycles of the two heads, the tape

(a)

(b)

Figure 13.3. Head drums. These photographs of the undersides of video head drums show the head chips protruding from the drum surfaces. In (a) is shown a Beta type by Sony - note the tacho magnet near the 'one-o'clock' position. In (b) appears a four-head drum for VHS format by Panasonic. Note the two windings (and two gaps) per chip

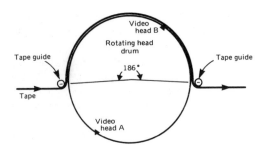

Figure 13.4. Omega wrap for a two-head drum. The tape occupies rather more than half a turn of the drum

wrap is in fact about 186°, slightly more than half a turn. This is known as an omega (Ω) wrap, outlined in *Figure 13.4.*

Track configuration

A typical track layout for a two-head helical VTR appears in *Figure 13.5.* Here we can see the video tracks slanting across the tape, shaded for head A, white for head B. At the edges of the tape, further tracks are present: the upper carrying a control track (serving a similar purpose to the sprockets in a cine film, and described later) and the lower carrying the sound track. Sound is recorded longitudinally in the same

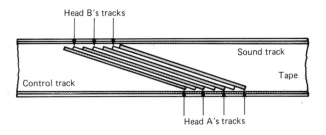

Figure 13.5. Video tracks written by a two-head system, with each head writing alternate tracks. The relative positions of the sound and control tracks are also shown (VHS format)

way as in an audio recorder, but with limited frequency response due to the low linear tape speed in the sorts of VTR we are dealing with. Budget machines only have this mono longitudinal sound track; other VTRs are additionally fitted for hi-fi sound. This superior system is described in Chapter 16. Tape is 12.65 mm (VHS) or 8 mm (Video 8) wide. It progresses at speeds between 23.4 and 10.06 mm/second, depending on format and mode.

Now let us examine the practicalities of the mechanical arrangements of helical scan machines.

Tape threading

One characteristic of all domestic VTRs is the storage and transport of the tape in an enclosed cassette. This not only prevents contamination of the tape and physical damage to it, but also makes easy the loading and operation of the machine. Unlike an audio cassette, where the capstan, pinch-wheel (pressure roller) and heads 'come to visit', as it were, the video cassette system requires that a loop of tape be drawn from the cassette and wrapped around the video head drum, stationary heads, tape guides and capstan assembly. This is called threading, and of the several ways of going about it, we will describe two.

VHS VTRs use, as we have seen, a co-planar cassette similar in form to the compact audio type, but larger. To load the machine, the cassette is fed horizontally into a carrier which is itself then pushed downwards into the machine. This action also opens the hinged front flap of the cassette. A pair of posts penetrate into the cassette shell (see *Figure 13.6a*) and when the thread mode is initiated they move away from the cassette, drawing out a loop of tape in 'M' formation. At the limit of their travel, the posts locate in 'V' notches mounted vertically at each side of the head drum to give the required 186° head wrap. This completes the threading operation, and the posts now form guide rollers on the tape path, as can be seen in *Figure 13.6b*.

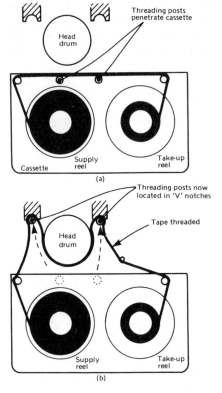

Figure 13.6. VHS threading system. (a) shows start of the threading and (b) threading completed

An alternative arrangement is used in many Betamax machines, and in some versions of VHS format recorders. Here (*Figure 13.7a*) a single post projects into the cassette when it is pushed home. The post is mounted on a *threading ring* which is motor driven during the thread mode. In our example, the ring is driven anticlockwise to thread in, pulling the post and tape loop with it. When the ring has completed almost three-quarters of a turn it stops, by which time the tape is wrapped around the video head drum and in position for normal transport to begin, as shown in *Figure 13.7b*. In this system the ring surrounds the head drum and stationary heads. A broadly similar threading system is used in the very compact Video-8 format machines, but the threading ring is very small, and 'tight' around the head drum.

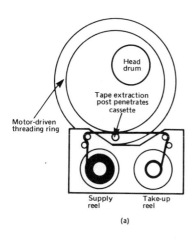

(a)

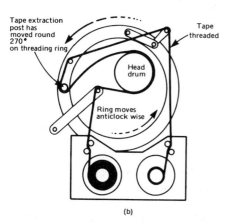

(b)

Figure 13.7. The Betamax threading system. (a) start of threading, and (b) threading completed

Tape path

For Beta and 'homebase' VHS VTRs, the mechanical and electrical components on the tape deck do not vary fundamentally between any of the formats or manufacturers. Let's follow the career of the tape on its involved journey from the supply spool to the take-up spool. *Figures 13.8* and

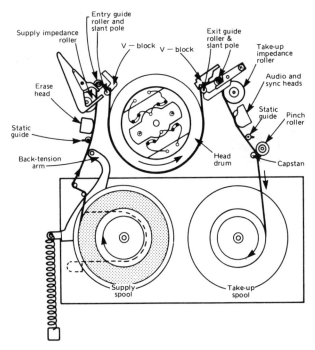

Figure 13.8. The path of the tape through the deck

13.9 show a typical deck, with the tape threaded and ready to go. The supply spool is on the left-hand side. On emergence from the cassette, the tape tension is checked by the tension arm, which brakes the supply spool to maintain tape tension constant. On most machines this is a purely mechanical operation, though sophisticated designs use electrical circuits in a *tension servo*. The tape next encounters the full-width erase head, which in record mode is energised with a high-frequency, high-amplitude CW signal to wipe clean all recorded tracks, video, sound and control. Its mode of operation is the same as the erase head of an audio

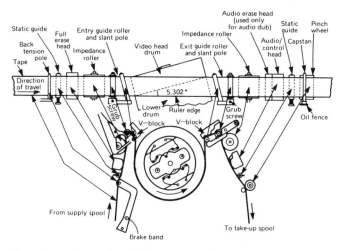

Figure 13.9. Tape path and 'linear' sequence of deck components

machine. From here the tape passes around a *supply impedance roller* to iron out any speed or angle fluctuations imparted by the tension arm or the tape's passage over the erase head. An entry guide roller, part of the threading assembly, passes the tape to the slant pole. The latter is slightly inclined from the vertical, and the effect of this is to create more tension on the top edge of the tape than at the bottom. The result is that the tape ribbon is 'nudged' downwards so that it sits firmly on the critically-positioned ruler edge around the video head drum.

The head drum itself is tilted at an angle of about 5° so that the head-wrap angle of the tape is correct with the tape moving parallel to the horizontal deck surface. On leaving the head drum the tape is again nudged downwards by an inclined exit guide to ensure that it is correctly seated on the ruler edge all the way round the 186° wrap of the head drum assembly. The tape now passes over the exit guide roller/ threading post and is steadied by a take-up impedance roller. Now the tape encounters the audio/sync head, a single assembly with separate heads lined up with the upper and lower edge of the tape. A further roller keeps the tape aligned on its way to the capstan, the prime mover of the tape transport system.

The capstan is a precision-machined shaft, motor-driven under the influence of a servo system and a stable timing reference. Holding the tape tightly in contact with the capstan is the pinch roller, which is disengaged during stop, pause, and fast transport modes. Finally the tape passes back into the cassette and on to the take-up spool, which is gently driven by a slipping clutch or direct-drive motor. The sequence of deck components is illustrated in the 'linear' diagram at the top of *Figure 13.9*. In a given format, all models by all manufacturers will conform to this 'linear' diagram so that tapes recorded on any machine will play back on any other. This compatibility is an essential feature, demanding that all machines agree with the parameters laid down in the format specification. The most critical and significant parameters are: video head positioning with respect to height and angular mounting on the head drum; head-to-ruler edge angle on the video head drum; drum exit to sound/sync head spacing; linear tape speed; and of course the track configuration, which we will shortly examine in more detail.

Other deck components

Before we leave the deck, however, there are several other devices to mention, mainly concerned with the *systems control* or safety aspect of the VTR. The most important of these are shown in *Figure 13.10*, starting with the end-of-tape sensors (AA). These prevent physical damage to the tape by inhibiting transport when the end of the tape is reached in

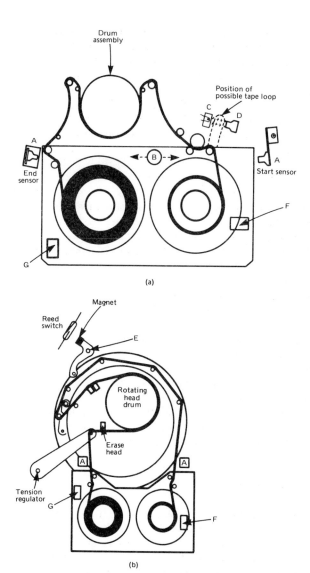

Figure 13.10. System control sensors on typical tape decks: (a) VHS; (b) Betamax

either direction, and permitting new motion only in the 'safe' direction. For VHS the sensors are optical, depending on a centrally-mounted filament bulb (early models) or infra-red LED emitter (late VHS and Video-8 machines), B, to illuminate photo-electric sensors through clear leader tape at each end of the tape. In some front-loading machines the passage of the cassette itself through these light paths is used in place of 'F' as a 'cassette-in' indicator. The Betamax machines have a rather more sophisti-cated end sensor. Here we find a metallic strip at each tape end, not to operate a contact system, but to inductively load down an oscillator coil mounted on the deck. The presence of the strip stalls the oscillator, invoking a stop signal.

Components C and D in *Figure 13.10a* together form a slack sensor. If the tape ribbon goes slack, either the tape has broken or transport has stopped, and in either case the machine must be stopped quickly. This is achieved by an illuminated LED (light-emitting diode) shining on the surface of a phototransistor. If a loop of tape comes between them, the phototransistor turns off, and stop mode is initiated. On some machines, the same slack-sensing effect is given by a tension arm E carrying a bar magnet, as depicted in *Figure 13.10b*. When the arm moves over, the magnet passes over a reed switch to invoke the stop mode.

Under the cassette carriage there are two microswitches, F and G. One is the 'cassette in' indicator which inhibits mechanical action in the absence of a cassette. The other is a 'tab detector' which prevents accidental erasure if the safety tab has been removed from the cassette in question. More details on these will be found in Chapter 18.

Scanning systems

In the original VTR plan, a guard band was left between video tracks on the tape. This was true of the first machine to appear on the domestic scene, the Philips N1500. Linear tape speed here was over 14 cm/second, and the track configuration is shown in *Figure 13.11*. It can be seen that each video track is spaced from its neighbours by an empty guard-band, so that if slight mistracking should occur, crosstalk between tracks could not take place. Each video track was 130 μm wide and the intervening guard bands 57 μm wide. This represents relatively low packing density of information on the tape, and it was soon realised that provided the tape itself was up to it,

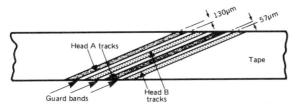

Figure 13.11. Guard-band recording: the track formation for VCR format as recorded by the Philips N1500 machine

a thinner head could be used to write narrow tracks; if the linear tape speed was also slowed down the tracks could be packed closer together, eliminating the guard band. Using both ideas, tape playing time for a given spool size could be doubled or trebled. First, though, the problem of crosstalk had to be solved. Even if the mechanical problems in the way of perfect tracking could be overcome so that each head always scanned down the middle of its intended track, crosstalk would occur due to the influence of adjacent tracks, and the effect on the reproduced picture would be intolerable. A solution to this problem was found in the form of azimuth recording.

The azimuth technique

For good reproduction from a tape system it is essential that the angle of the head gap on replay is exactly the same as was present on record, with respect to the plane in which the tape is moving. In an audio system the head gap is exactly vertical and at 90° to the direction of tape travel. If either the record or replay head gap is tilted away from the vertical,

233

even by a very small amount, tremendous signal losses occur at high and medium frequencies, the cut-off point travelling further down the frequency spectrum as the head tilt or *azimuth error* is increased. If the same head is used for record and replay (as is usually the case in audio tape recorders) the azimuth error will not be noticed, because there is no azimuth difference between record and replay systems. A pre-recorded tape from Granny in Scotland will not be up to much, however!

This azimuth loss effect, bad as it is for Granny, is the key to successful recording and replay of video signals without a guard band. Let's designate our video heads A and B, and skew A's head gap 15° clockwise and B's head gap 15° anticlockwise as in *Figure 13.12a*. This imparts a total 30° difference in azimuth angle between the two heads, and the result is video tracks on the tape like those in *Figure 13.12b*.

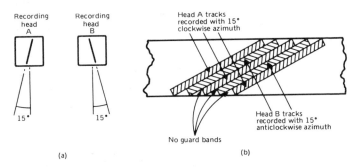

Figure 13.12. Azimuth offset. The head gaps are cut with complementary azimuth angles so that the guard bands of *Figure 13.11* can be eliminated

With the built-in error of 30°, head A will read virtually none of head B's tracks, and therefore the guard band can be eliminated. This was the *modus operandi* of the VCR-LP format, using the same cassette and virtually the same deck layout as the original VCR format, but with linear tape speed reduced by 50 per cent, and video track width down to 85 μm. It worked, and the two-hour machine was a reality.

Subsequent formats use a smaller azimuth tilt: 6° for VHS, 7° for Betamax and 10° for Video 8. The offset *between* heads is double this figure in each case.

Compatibility

The format specification sets out the parameters of the recorded signal on the tape and it is important that each machine in record and playback conforms exactly to these parameters. If a VTR is in a worn or maladjusted state, electrically or mechanically, it may well record and play back its own tapes satisfactorily, but mistracking and other problems will arise when it is called upon to replay tapes from another machine, or when its tapes are replayed elsewhere.. Possibly the most common cause of incompatibility in a VTR is mechanical misalignment of deck components, particularly head-drum entry/exit guides, dirt build-up on the head drum ruler-edge guide and physical displacement of stationary heads and pinch roller.

To check mechanical and electrical alignment a precision test tape is used, recorded on a 'design centre' machine under carefully controlled conditions. Also required is a series of mechanical jigs produced by the VTR manufacturer, as described in Chapter 20. A properly aligned machine will be compatible with others of the same format, needing only an adjustment of the tracking control during replay. The function and need for this control will be described in Chapter 17.

Television recap

Before we go on to describe the laying-down of the video tracks in detail, let us briefly recap on the characteristics of the UK television system, known as CCIR System I. For the purposes of this chapter we need only examine the luminance, or monochrome component, of the signal; the colouring signals will come under scrutiny later!

As is well known, the picture is made up of 625 lines, each drawn from left to right of the screen as we view it. For bandwidth conservation reasons (see page 3) the lines are not transmitted in sequence, but in interlaced fashion. This means that line 1 is traced out at the top of the screen, then a gap is left before drawing line 2. Below line 2 another gap is left between it and line 3, and so on, all the way down to the bottom of the TV screen. By the middle of line 313, we are the centre of the bottom of the screen, and at this point the scanning process is suddenly terminated, recommencing at the top of the screen. So far we have traced out one *field* of 312½ lines in a period of 1/50 second or 20 ms. During this time the instantaneous brightness of the scanning spot along each line has been changing in sympathy with the video signal to build up the picture. The scanning lines for the first field are shown in solid line in *Figure 13.13*.

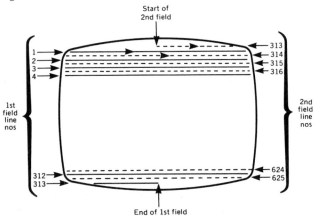

Figure 13.13. Interlaced TV scanning, in which the lines of field 2 are traced in the gaps between those of field 1

The lines of the second field (nos 313 to 625) are slotted into the gaps between the lines of the first field as the vertical timebase commences its second journey down the screen, shown by the dotted lines of *Figure 13.13*. The second field lasts another 20 ms, and the two fields combined make up a television *frame*, one of which is completed every ⅟₂₅ of a second, or at 40 ms intervals. Each field contains 312½ lines, and the duration of each line is 64 μs.

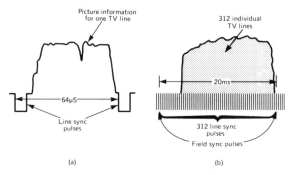

Figure 13.14 Line and field synchronising pulses in the television waveform

The television waveform contains two sorts of synchronising pulses, one at 64 μs intervals to define the starting point of a new line and a more complex one at 20 ms intervals to signify the beginning of a new field. These are illustrated in *Figure 13.14a* and *b* respectively, and are separated from the vision signal in the TV receiver to initiate the flyback or retrace stroke of the line and field timebases. Thus the video signal is 'chopped' as it were, at line and field rate, and a sharply-defined pulse inserted as a timing reference. In a VTR the timing reference at field rate is a useful marker, and is used to set the video head position and define the length of a video track on the tape. We can now relate the TV picture to the magnetic 'signature' it writes on the video tape.

The video track

Figure 13.15 shows four adjacent tracks on the tape. Track 1 is laid down by head A, and the 'phasing' of the spinning video head drum is arranged so that the A head enters on to the

237

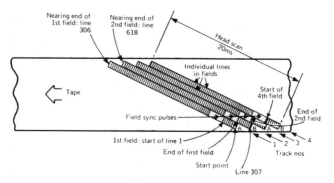

Figure 13.15 The positions of TV lines and fields in the recorded tape track. Each head writes one field of video information

tape and starts to record just before a field sync pulse arrives. It will write about seven lines of picture before recording the field sync pulse, and then go on to write the rest of the lines in the field. By the time line 306 has been recorded, head A is leaving the top of the tape, having recorded one field of 312½ lines; head B has entered onto the tape and is about to record track 2, consisting of the next sync pulse and field; and the head drum has turned through half a revolution, or 180°. During this time the capstan has pulled the tape through the machine just far enough to ensure that track 2 lays alongside, and just touching, track 1. Track 3 is laid down by head A again, track 4 by head B, and so on.

We can see, then, that signals recorded towards the lower edge of the tape correspond to those in the top half of the picture and vice-versa. Thus a tape damaged by creasing along the top edge may be expected to give a horizontal band of disturbance in the bottom half of the reproduced picture. Around the period of the head changeover point, both heads are at work for a brief instant, one just about to run off the top of the tape, and the other having just entered at the bottom. Thus there is an overlap of information. All VTRs incorporate a head switch which electronically switches between the heads at the appropriate time, just before the field sync pulse. As should now be clear, this takes place at the very bottom of the

Figure 13.16   The head changeover at the bottom of the TV picture. It is normally hidden, and the picture height has been maladjusted here to show it, appearing as a 'tearing' effect on the lower third of the bottom castellations in the test card

picture, and any picture disturbance due to head changeover during record and replay is hidden by the slight vertical over-scanning which takes place in a correctly-adjusted TV. *Figure 13.16* shows the head changeover on a TV whose picture height has been reduced to demonstrate the effect.

Still-frame considerations

In domestic VTRs the video head drum and the tape move in the same direction, so that with the head drum spinning anticlockwise (as they do in all formats) the tape is pulled through the half-wrap round the head drum in an anticlockwise sense. This, in conjunction with a special lapped surface on the rim of the head drum, ensures a minimum of friction between tape and drum.

From our studies so far, and the video track diagrams already described, it might reasonably be supposed that if the tape transport were stopped while video head rotation continued, the heads would repeatedly scan the same track over and over again to produce a good still-frame picture. In fact it does not happen, for reasons which will become clear.

Referring again to *Figure 13.9*, the angle between the plane of the head path and the ruler edge around the head-drum mounting is 5.302° (VHS system) and when the tape is stationary the video heads move across the tape at this precise angle. Once the tape is moving, however, the tape and head velocities subtract, because the head and tape are moving in the *same* direction. Think of the tape as a slow lorry being overtaken by a fast car representing the video head; the car's speed *relative to the lorry* is less than if the lorry were stationary. This effective slowing of the writing speed means that in the fixed time available (20 ms, one half-turn of the head drum) the track angle will be steeper than the 5.302° set by the guide, and the recorded track length shorter than when the tape is stationary. The relative track angles and lengths for stationary and moving tape are shown in *Figure 13.17*, where it can be seen that the track angle has changed to 5.328° when the tape is moving in record and replay. Although the angles and parameters quoted are for VHS, the same principle applies to all formats. Let's see what effect this has on still-frame reproduction.

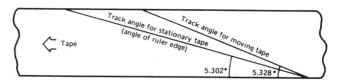

Figure 13.17 The effective video track angle changes when tape transport is stopped for reasons explained in the text

240

Figure 13.18. Head-mistracking effect on playback picture

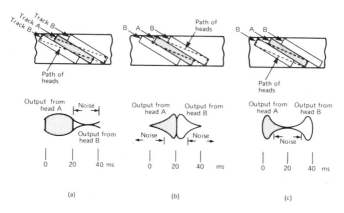

Figure 13.19. The position of the mistracking bar depends on the exact point, relative to the heads' path, at which the tape stops

241

The video track, recorded at an angle of 5.328°, is now stationary and being scanned by a head moving across the tape at 5.302°. The difference is not very great, but with video track width of the order of 49 μm it is sufficient to cause the head to diverge from its intended track on to an adjacent one, which, as we have already seen, is recorded with the 'wrong' azimuth angle. This results in mistracking and a band of noise on the reproduced picture, as shown in *Figure 13.18*. Where the noise band occurs depends entirely on the precise stopping point of the tape. *Figure 13.19* shows three possibilities, with their effects on the still-frame picture. In simple VTRs where a stop-motion facility is provided, this unavoidable noise band precludes any serious use of the feature, and it is referred to as 'picture pause' rather than 'still frame'.

Picture freeze

Second-generation VTRs have special facilities for the display of noise-free still frames. First, let's dismiss what appears to be the simple solution, that of altering the angle of the head ruler edge during still-frame replay. It would appear that slightly tilting the ruler edge or inclined tape guide around the head drum mounting would realign the tracks with the head path during still-frame reproduction, and solve the problem. The guide is precision-machined into the lower head drum with great accuracy, however, and cannot be altered. Even if the inclined tape guide could be somehow made to move slightly on the lower drum, it would be mechanically impossible to arrange a system whereby it was physically tilted through the tiny angle involved with any useful degree of accuracy and repeatability. Other solutions have to be found!

One approach is to make the video heads wider, so that in the stop-motion mode the replay head is 'reading' a sufficiently broad path to embrace the change in track angle without undue loss of signal. The idea is shown in *Figure 13.20* where a head 59 μm wide is able to keep in sight the 49 μm video track throughout the field period. It must be appreciated that in stop motion both heads trace the *same path* across the tape, so that if track 1 (head A's province) is being scanned, head B will produce no output from it due to the azimuth error (refer again to *Figure 13.19*). To obtain a useful

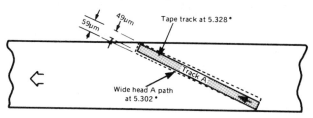

Figure 13.20. A special wide video head scanning a video tape track during 'stop motion'

output from head B, then, it needs to see some of its own track 2, and to this end it is made even wider than head A, in fact 79 μm. Even when travelling along the same path as head A, then, it reads enough of the adjacent track 2 to provide a usable output, though the picture quality in still frame is, not surprisingly, inferior to that on normal replay. The extra-wide heads are also necessarily used during record to lay down normal compatible 49 μm tracks according to the format

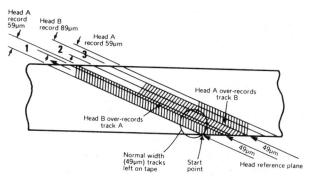

Figure 13.21. Laying down standard-width tracks with wide recording heads. The excess track width is 'wiped off' by the next head sweep

243

specification. This is achieved because the lower edge of both heads are on the same reference plane, and each new track recorded by either head will wipe off, or over-record, the excess width of its predecessor, as shown in *Figure 13.21*.

Three-head and four-head drums were introduced to improve still-frame reproduction. The extra heads are wide, and optimised for 'freeze' reproduction; a further advantage was that both 'trick' heads could now be cut with an azimuth angle to scan *the same track*. *Figure 13.22* shows one such approach. Electronic field-store memories are now used for consummate freeze-frame pictures in home VTRs.

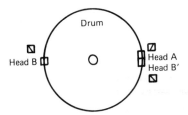

Figure 13.22. One form of three-head drum. In record and normal playback heads A and B are used. For still-frame, heads B and B¹ scan the *same video track* for noise-free and jitter-free reproduction

Miniature VHS head drum

The 62.5 mm drum diameter of the standard VHS specification is a great handicap in portable video equipment. To achieve a deck size small enough to be accommodated in a light camcorder a small head drum is used: it is 41.3 mm in diameter. To permit record and playback of *standard* VHS tracks some complexity in the mechanics and electronics of the machine is unavoidable, and *Figure 13.23* shows the essence of the arrangement. The travel of the tape guides follows a longer and more sinuous path than before, to wrap the tape around 270° of the periphery of the small drum, which rotates at 2250 r.p.m. This high speed is calculated to sweep a single drum-mounted head along the entire length of a *standard* VHS video track during its contact with the wrapped tape. The inclined and continuous ruler-edge around the lower drum assembly maintains the tape at the normal 5.302° angle to the head-sweep path.

Plainly, one pair of heads will not suffice to work this system. At the point when one head is leaving the tape wrap, there needs to be another just 90° ahead, the point where it is just entering the tape wrap. This ensures continuity of signal feed onto the tape, whose linear speed around the drum conforms to standard VHS specifications – 2.34 cm/s for SP mode, 1.17 cm/s for LP mode. If the head which has just left the tape is writing or reading 'A' tracks the one ahead of it and the one behind it must be 'B' heads, with azimuth angles cut accordingly. Hence the A-B-A-B configuration of the four heads around the drum in *Figure 13.23*. Each head scans every *fourth* track on the format-standard tape.

At any given moment only one of the four heads will be active in record or playback, and since two others will be in contact with the tape at this time a four-phase head-switching system is

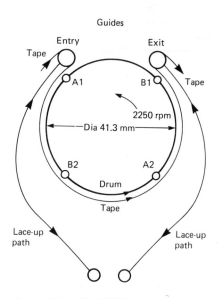

Figure 13.23. Small VHS head-drum. Four heads are required to read and write standard tracks

245

required during both record and replay. *Figure 13.24* shows the switching sequence and the time relationship of the video signal to the active period of each head. The switching system is the same in record and playback modes, though of course the routing of the video FM carrier is opposite.

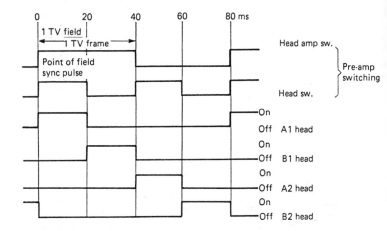

Figure 13.24. Sequential switching for the heads in *Figure 13.23*. Switching is applied on record and playback

It can be seen, then, that four heads are necessary in a 'small' VHS head drum to do the work of the two in a conventional drum. If separate heads are provided for SP and LP eight heads are required on the drum, though they can be mounted in four chips, each carrying two windings and two head gaps. Some VHS camcorders are additionally fitted with a flying erase head, giving an effective total of *nine* heads around the drum periphery, with a multi-winding rotary transformer to couple the recording signals to the heads.

Video-8 format was designed from the outset for a two-head 40 mm-diameter drum, so these complications do not arise.

Tracking

We have seen that during record the phasing of the head drum position relative to the incoming video ensures that the field synchronising pulse is laid down at the beginning of each head scan of the tape. During replay it is necessary to set and maintain the relative positions of the tape and head drum so that each head scans down the centre of its own track. The record phasing and replay tracking is carried out by the VTR's servo systems, which have a chapter to themselves later in this book. Suffice it here to say that on record a *control track* is recorded along the edge of the tape as a timing reference. This takes the form of a 25 Hz pulse train and is used on replay to set and maintain the relative position of the tape tracks and head drum.

Automatic tracking systems

The system of marking the physical position of each video track on tape by control-track signals is a long-established one, and lends itself to various methods of quick programme finding by modifying the control track pulses temporarily at the beginning of each new recording. The VISS and VASS search systems used in VHS machines detect the change of control-track pulse formation in search, FF and REW modes to stop transport.

Servo control by a separate tape track, however, requires provision and careful alignment of a separate pulse record/replay head; and a tracking control which in many cases is a manual (user-operated) type. Correct tracking with this arrangement is also dependent on correct tape tension and precision alignment of tape guides. For these reasons and others, control-track systems are vulnerable to tracking errors, especially in LP modes where the video tracks on tape are narrow and less tolerant of head-path errors.

Automatic tracking finding

Automatic track finding (ATF) is a more advanced form of tracking, in which head-guidance signals are continuously recorded *in the video tracks*. The concept was first introduced in the now-defunct Philips/Grundig V2000 format, where it was known as DTF (dynamic track following) and was used with a piezo-bar mounting system for the heads, whose position could be set by applying DC deflection voltages to the piezo bars.

Video-8 format uses an ATF system, the essence of which is a pilot tone which is recorded with the picture throughout every video track. Four pilot tone frequencies are used: f1, 101.02 kHz; f2, 117.19 kHz; f3, 162.76 kHz; and f4, 146.48 kHz. They are added to the luminance FM record signal and recorded in the sequence f1, f2, f3, f4 in successive head sweeps, see *Figure 13.25*. Relatively low frequencies like these are almost unaffected

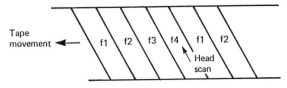

Tape movement ← f1 f2 f3 f4 f1 f2

Head scan

Figure 13.25 Tone sequence laid on tape for ATF

by any azimuth offset of the replay head, so pilot-tone crosstalk from adjacent tracks is easily picked up by the video heads during playback. The pilot-tone frequencies are chosen to have specific relationships as *Figure 13.26* shows. The beat frequencies which arise when pilot tones from adjacent tracks are mixed are always 16 kHz or 45 kHz. These beat products are used to steer the head path/tape track alignment for optimum tracking: when the levels of 16 kHz and 45 kHz beat signal are equal the replay head must be scanning along the dead centre of its video track, indicating optimum tracking.

There are several ways in which the ATF pilot tones can be processed during playback. A simple one, illustrated in *Figure 13.27*, utilises two bandpass filters to pick off and separate the 16 kHz and 45 kHz beat products so that they can be separately detected and measured. The DC outputs are applied to the differential inputs of an operational amplifier whose output forms the error signal. This error output can be used to phase-lock either

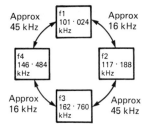

Figure 13.26. ATF tones have a carefully-selected frequency relationship

the capstan or head-drum servo to give accurate and continuous tracking correction with built-in compensation for 'mechanical' errors, tape-stretching etc.

ATF playback

While *Figure 13.27* shows a simplistic approach to replay ATF processing, the actual system used is more sophisticated. Pilot tones are generated during playback, again changing on a track-sequential basis triggered by head tacho pulses. This time the tones are sequenced in reverse order, however: see *Figure*

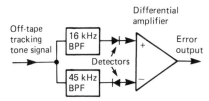

Figure 13.27. Basic arrangement for derivation of an error voltage from inter-track pilot beats

13.28. The top waveform represents the head-drum flip-flop signal, high for head A, low for head B. The off-tape pilot tone (REC pilot) is shown below, and at the bottom the newly-

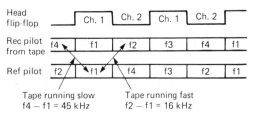

Figure 13.28. Replay beat-tone generation: off-tape pilot frequencies are compared with a locally-generated tone sequence

generated local pilot tone (REF pilot) in the reverse order f4, f3, f2, f1. The diagram shows ideal tracking conditions, in which the off-tape tones switch in synchronism with the REF pilot tones. At every fourth field two f1 tones appear simultaneously at a mixer, whose output consequently drops to zero. During the other (properly-tracked) fields the beat product is either zero (f3/f3) or 29 kHz (f2/f4), the latter being outside the band of interest and thus rejected.

If the tape speeds up – see *Figures 13.25* and *13.28* – the REC pilot moves to the left, permitting REC f2 to appear during the period of REF f1 and giving rise to a 16 kHz beat product. As other tracks are scanned all the beat products (f3/f4, f4/f3, f1/f2) are 16 kHz.

Conversely, when the tape slows down the REC pilot pattern moves to the right. Some REC f4 now appears during the REF f1 period to produce a 45 kHz beat product. Similarly during subsequent scans REC f1 beats with REF f4, f2 with f3 and f3 with f2, producing a 45 kHz beat output in each case.

Thus a fast-running tape always results in a 16 kHz output from the mixer; and a slow-running tape always produces a 45 kHz output. By using suitably tuned bandpass filters to select these products, and feeding their outputs to separate peak-detectors an ATF error signal is produced for subsequent smoothing and passage to the capstan servo control input. In practice the replay ATF processing circuit is also provided with artifices to detect 'false lock' conditions and to permit locking of mistracking-noise bars during search modes.

250

Video-8 tape-signal spectrum

The DTF pilot tones are recorded at the lowest part of the frequency spectrum as shown on the left of *Figure 13.29*. Here they do not interfere with the 'signal' components of the tape recording. The other parts of the tape-signal spectrum will be dealt with later: luminance in Chapter 14, chroma in Chapter 15, and audio in Chapter 16.

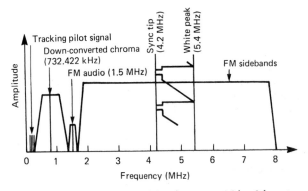

Figure 13.29.   The spectrum of signals on tape - Video-8 format

14

Signal processing: luminance

In previous chapters, we have seen some of the problems of tape recording video signals, and have examined how these are overcome by the use of high writing speeds and FM modulation. We have seen, too, how home VTRs achieve an acceptable performance at low cost by signal bandwidth restriction. This is by no means the only shortcoming of the domestic machine! In spite of these, the playback picture from a domestic VTR in good condition is very acceptable; in this chapter and the next we will examine not only the basic circuitry used to process TV signals within the VCR, but also the special circuits and processes, some of which owe their origins to broadcast technology, which compensate for the deficiencies inherent in a cassette VTR system, and enhance performance.

The luminance (or black-and-white) signal is dealt with separately from the chrominance (or colouring) signals in home VTRs. For luminance the basic idea is to modulate the

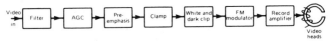

Figure 14.1. The stages in the luminance-recording process

signal on to an FM carrier for application to the recording head, and demodulate it to baseband during the replay process. As the simplified block diagram of *Figure 14.1* shows, however, there are several other processes undergone by the luminance signal, and these will be described in turn.

AGC (automatic gain control) and bandwidth limiting

The basic luminance signal that we wish to record may come from a TV camera or other local video source, or more likely, a broadcast receiver built into the VTR. In either case it will be positive-going for white and will probably contain a chrominance signal modulated on to a 4.43 MHz carrier. It is important that the signal recorded on the tape is within the limits of the recording system, so the luminance signal is first passed through an AGC amplifier with a sufficiently wide range to compensate for signal inputs of varying amplitudes. This works in a similar manner to the AGC system of a radio or TV, by sampling its output level to produce a DC control potential, and applying this to an attenuator at the amplifier input. Thus the output from the AGC stage will be at constant (say) lv amplitude.

In home VTRs, luminance bandwidth is restricted to about 3 MHz, give or take a hundred kHz or so between the formats. If higher frequencies than this are allowed to reach the modulator they will make mischief with sidebands, as explained in Chapter 12. A low-pass filter with a quite sharp cut-off around 3 MHz is incorporated in the record signal path, then, and this also eliminates all the chrominance components of the signal, which are based on a subcarrier of 4.43 MHz. When recording in monochrome more bandwidth

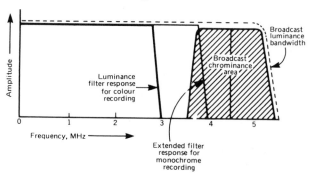

Figure 14.2. VTR luminance-recording filter characteristics relative to the spectrum of the composite video signal as broadcast

253

can be allowed, occupying the space normally reserved for the chrominance signal, and many machines have an automatically-switched filter characteristic for monochrome and colour recordings, as shown in *Figure 14.2*.

Pre-emphasis

Chapter 12 showed how pre-emphasis is used in an FM system to reduce noise. What is required is a boosting of high-frequency (HF) signals before the FM modulator, and this is achieved by a circuit like that in *Figure 14.3a*. Here we have a

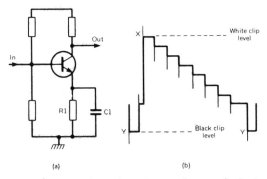

(a)   (b)

Figure 14.3.   (a) shows the rudiments of pre-emphasis circuit, and (b) its effect on the luminance staircase waveform

common-emitter transistor with a small capacitor (C1) in its emitter circuit. At high frequencies, the capacitor's reactance will become comparable with the emitter resistor R1, and the negative feedback due to the latter component will diminish, resulting in greater output from the stage (i.e. at the collector) at those frequencies. A typical pre-emphasis curve is shown in *Figure 14.4a*, in which 10 dB of pre-emphasis is given to frequencies above 1 MHz.

In some machines, particularly Video-8 and VHS-LP types (both make use of very narrow video tracks) a further, *non-linear*, pre-emphasis circuit is used. This applies a degree of HF boost which is dependent on signal level (greater for small signals, less for large signals). A family of curves for this system is given in *Figure 14.4b*. Although this calls for more complex circuits (the de-emphasis characteristic has to be non-linear to compensate) it does offer a useful reduction in noise level on replay.

Because the steep rising and falling edges of a luminance transient, or step, look like a high frequency they are effectively differentiated in the pre-emphasis process, so that

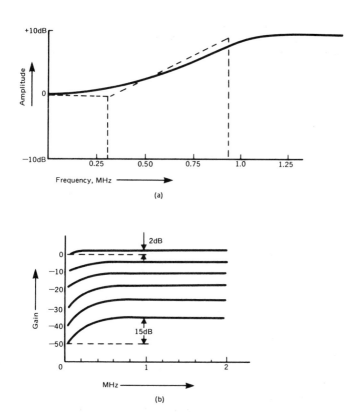

Figure 14.4. A typical VTR pre-emphasis curve is shown at (a). In (b) appears a family of curves for a non-linear pre-emphasis system, in which the degree of HF 'lift' varies from 2 to 15dB, depending on signal level

the 'risers' of a luminance staircase waveform take on the spiky characteristic shown in *Figure 14.3b*.

Where the spikes exceed peak white level as defined by the output of the AGC stage, we are once again in danger of driving the FM modulator into excessive deviation, and this must be prevented.

White and dark clip

The next process, then, is a clipping action on the video signal. The danger areas for overmodulation are those shown at points X and Y in *Figure 14.3b*, and the dotted lines indicate the levels at which the signal must be clipped. This action takes place in a circuit like the one of *Figure 14.5*, where diodes provide the clipping effect at both extremities of the

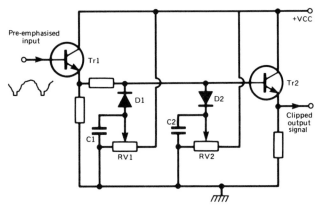

Figure 14.5. A diode threshold clipper; clipping levels are set by RV1 and RV2

luminance signal. Let's assume the sync-tip of a normal luminance waveform drives Tr1's emitter down to +1 V, and peak white takes it up to +4 V. If RV1 is adjusted so that +1 V is present at D1 anode, then ignoring the voltage drop across the diode itself, D1 will conduct on any luminance signal excursion below 1 V and the signal will be grounded in C1.

Similarly, RV2 is set to give +4 v at D2 cathode so that any luminance signal above this level turns on D2 and is grounded in C2. Thus RV1 becomes the 'dark clip' level control and RV2 the white clip level control. In practice, a little 'headroom' is permitted above and below the normal signal excursion before the clippers come into operation; some machines have only a white clip facility.

For this clipping system to work, the voltage levels corresponding to black and permissible peak white have to be closely defined. Because the AGC stage defines the overall signal amplitude, it is only necessary to clamp the black level to a fixed voltage to achieve this, and clamping is carried out immediately prior to the clipper stage, usually in a driven line-rate clamp similar to those used in TV receivers.

FM modulator

The modulator is one of the most critical stages in a VTR. Its output needs to have equal mark-space radio regardless of deviation, and deviation must be linear with respect to

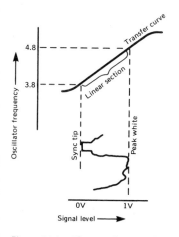

Figure 14.6. The transfer curve for an FM modulator. It has to be very linear between the frequencies corresponding to sync-tip and peak white. For the VHS system illustrated, these are 3.8 to 4.8 MHz

257

modulating (video) voltage. The output must not contain harmonics which would generate harmful sidebands, and frequency stability has to be of a high order. A constant output level is also required over the whole of the deviation range, but this can be achieved by limiting stages after modulation, as we shall see. Current VTRs use an IC modulator, but to explain the operation of the circuit we shall take as a model the astable, or multivibrator, type as used in early VTR designs. The curve of *Figure 14.6* shows the relationship between modulating voltage and output frequency for a typical system.

Astable modulator

The astable oscillator is a well-known configuration and is drawn in a form suitable for VTR use in *Figure 14.7*. The square-wave output is produced by alternate conduction in Tr1 and Tr2, the conduction period for each transistor being set by the capacitor/resistor (CR) combination in its base circuit. Thus C1/R1 determine the conduction period of Tr1; when Tr1 switches off Tr2 comes on for a period determined by C2/R2, and so on. The time-constant of these CR pairs is chosen to be identical, i.e. C1=C2 and R1=R2 so that the *mark-space ratio* of the output waveform is equal or 1:1. Now the conduction period for each transistor is also dependent on the *aiming voltage* of each of the timing capacitors, that is the voltage towards which it is charging via its timing resistor. If this aiming voltage is varied, the charging *rate* of the cross-coupling capacitors C1 and C2 varies in sympathy, and it is the charging *rate* of the capacitors which determines the output frequency. In the circuit of *Figure 14.7* the aiming voltage is set by the potential divider R3/R4, but this can be modified by an externally applied voltage at point X. Because this is common to both CR timers, the mark-space ratio will remain constant while the basic frequency is varied in proportion to the voltage applied to point X. This, then, is how voltage variations (the video signal) are converted to frequency variations in the output of the astable modulator.

We have seen that harmonics are harmful in the output from an FM modulator, and the square-wave output characteristic of an astable circuit is very rich in harmonics! These are lost in the tuned and balanced transformer T1, whose output is virtually sinusoidal.

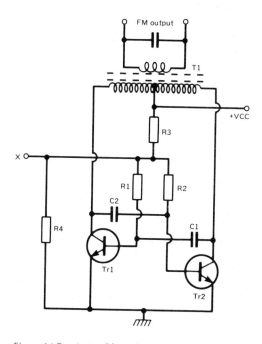

Figure 14.7. An astable multivibrator for luminace FM modulation

As frequency deviation is proportional to applied voltage, the carrier frequency and deviation are set by adjusting the video DC level and gain in the preceding circuits. Thus in our previous examples, basic carrier frequency (corresponding to black level) would be set by potentiometer adjustment of the video clamp voltage, and maximum deviation by presetting the operating point, or gain, of the AGC amplifier.

259

Limiting and head drive

A degree of amplitude modulation can occur in the FM conversion and coupling processes, and it is important that this is ironed out before the signal is applied to the recording heads. Typically, diode clippers are used in a circuit akin to that already given for white and dark clippers. The difference here is that the clipping action is constant, a process known as limiting. In *Figure 14.8* the diodes D1 and D2 limit the signal to a constant level for application to the recording amplifier.

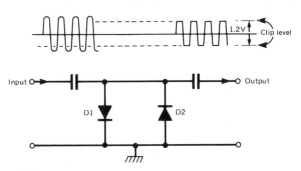

Figure 14.8. Simple diode amplitude clipper

The recording amplifier has two basic functions: to provide power amplification to drive the recording heads, and to apply recording equalisation to compensate for the falling response of the video heads and head/tape interface with increasing frequency. The circuit diagram of *Figure 14.9* shows the arrangement. The first stage after the limiter, Tr1, provides equalisation by giving a degree of lift to the higher frequencies. Frequency-selective negative feedback is again used, the operative component being C1. An emitter follower stage Tr2 acts as a buffer to drive the complementary output pair Tr3/Tr4. These work in similar fashion to audio and field output stages in radio and TV sets, but at a much higher frequency.

The level of the FM luminance signal fed to the recording heads is critical, and is set at an optimum level for the type of

260

videotape in use. The constant-level carrier acts as a recording bias signal for the colour signal which accompanies it on to the tape, and the critical FM luminance *writing current* is set by VR1 in *Figure 14.9* to bias the recording level to the centre of the linear flanks of the transfer characteristic shown in *Figure 12.4*.

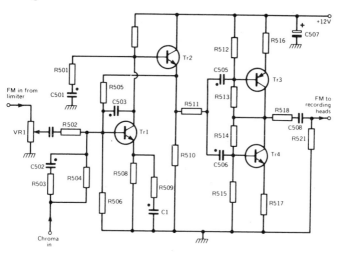

Figure 14.9. An FM recording amplifier by National Panasonic

Transferring the video signal to a pair of heads rotating at 1500 rpm is something of a problem, and the later recovery of the tiny playback signal an even greater one! Any form of brushgear is impractical from the noise point of view, so rotating transformers are used, with the primary winding stationary and the secondary winding rotating with the heads. These roles are reversed during playback, of course. Two separate and independent transformers are used, having printed windings and ferrite cores; signal transfer takes place across a very narrow air-gap between the two ferrite discs. Multi-head VTR designs use separate and concentric transformers. Except for small-drum VHS systems (Chapter 13) head switching is not carried out on record, both heads being driven together via the rotary transformers from the output of our push-pull recording output stage (*Figure 14.9*).

Replay circuits

During replay, the off-tape signal from the video heads is very small, and to maintain the necessary $> 40\,dB$ signal-to-noise ratio in the reproduced picture, low-noise amplification is necessary. The main replay processes after the preamplifiers are head switching, equalisation, limiting, drop-out compensation, demodulation to baseband, de-emphasis, crispening and amplification, after which the luminance

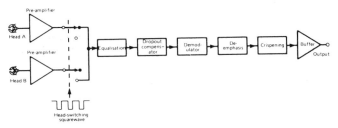

Figure 14.10. The replay chain for the luminance signal

signal is restored to its original form, usually 1 v peak-to-peak, negative-going syncs. A block diagram showing the order of the replay circuits is given in *Figure 14.10*. We will examine these in turn.

Head preamplifier and switching

The RF signal from the rotary transformers contains signal components ranging from 100 kHz to over 8 MHz. These are at high impedance and very vulnerable to noise pick-up. A typical pre-amplifier circuit appears in *Figure 14.11*. The replay signal is applied to a balanced transformer T1 whose windings resonate with trimmer C1, set to 'peak' the LC circuit at about 5 MHz. This resonance is damped by variable resistor RV1 to provide a reasonably smooth frequency response at the input to Tr1. An FET device is used here, exploiting its characteristics of high input impedance (to minimise loading of the head circuit) and low noise figure.

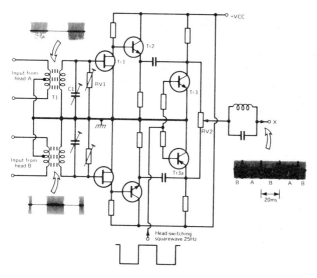

Figure 14.11. Head pre-amplifier circuit. Tr3 and Tr3a are head-switching transistors, and RV2 is a balance control to equalise the signals from each video head

Although both heads are driven together during the record process, on all current formats head switching is carried out on replay. This serves the dual purpose of eliminating noise from the inoperative head (remember only one of the pair is scanning the tape at any given moment) and sharply defining the head changeover point to minimise picture disturbance during changeover. In *Figure 14.11* head switching is carried out by Tr3 and Tr3a. An incoming square-wave signal at 25 Hz derived from a *tachogenerator* on the head drum switches Tr3 and Tr3a alternately into saturation, shorting to ground the output of each head during its 'passive' period. The FM replay signal at point X is continuous, then, and sourced from each head alternately for 20 ms periods.

Equalisation

Chapter 12 explained why during replay the signal from the tape head falls at the rate of 6 dB per octave. HF losses due to the approach of Fex are largely compensated for by the resonance circuit associated with the replay head and described above. To even-out the response on replay, further equalisation is provided, and in practice it takes the form of a 'boost' in the 2–3 MHz region, the major area for the lower FM sideband of the luminance signal. This is often catered for by the provision of an inductive component in the collector circuit of the preamplifier or a following stage.

Limiting

For correct operation of any FM demodulator, it is important that its input signal contains no amplitude variations. For this reason, several stages of limiting are usually provided in the path of the FM replay signal to clip it to a constant level. The form of diode clipper shown in *Figure 14.8* (D1 and D2) is suitable for this. In current VTR designs, limiting normally takes place (along with much of the other replay processing) within an IC.

Dropout compensation (DOC)

A videotape is not the perfect medium that we would like it to be! The magnetic coating is not completely homogenous, and with video track widths smaller than the diameter of a human hair even a microscopic blemish in the magnetic coating will delete some picture information. As the tape ages, slight contamination by dust and metallic particles, and oxide-shedding effects, will aggravate the situation. The effect of these tiny blemishes is a momentary loss of replay signal known as a *dropout*. Unless dropouts can be 'masked' in some way a disturbing effect will take place in the form of little ragged black or noisy 'holes' in the reproduced picture.

In practice, the video information on one TV line is usually very much like that on the preceding line; so that if we can arrange to fill in any dropout 'holes' with the video signal from the corresponding section of the previous line, the patching job will pass unnoticed. What's required, then, is a delay line capable of storing just one TV line of 64 μs duration, so that whenever a dropout occurs we can switch to the video signal from the previous line until it has passed. For delay-line bandwidth reasons, this is difficult to achieve at video baseband frequencies, so it is carried out on the FM signal before demodulation. To avoid disturbance on the picture the switching has to be very fast, virtually at picture-element rate. One form of DOC circuit is shown in block-diagram form in *Figure 14.12*. The FM replay signal takes three

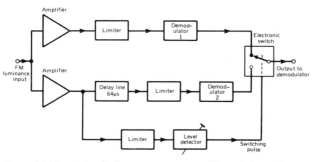

Figure 14.12. A simple dropout compensator. Demodulator 2 is operating on an FM signal one TV line earlier than that in demodulator 1

paths, the upper of which is via demodulator 1 and on through the following replay circuits; this is the path taken by the signal under normal circumstances, that is, when no dropouts are present. The bottom path consists of a level detector which monitors the FM signal from the playback heads. When a dropout comes along, the FM signal falls

265

below the detector's threshold level (normally pre-set by a potentiometer) and the detector output falls to zero. This operates a fast diode changeover switch S1, which now selects the middle path via demodulator 2, whose FM input is exactly 64 μs (one TV line period) 'old' due to the delay line, and thus corresponds to the FM replay signal at the same point on the previous line. This signal is held by the switch until the dropout has passed and full FM input is restored at the DOC input, when the detector output reappears and the changeover switch drops back to the output of demodulator 1.

Recycling DOCs

The system outlined above breaks down when any dropout exceeds one TV line duration, and this happens often – the tape area occupied by one line is microscopic. If the dropout period exceeds one line, then, both demodulators in *Figure 14.12* will be looking at noise, and a disturbance will be visible on the TV screen. To prevent this, further steps must be taken.

In early designs, a 'grey oscillator' was used, consisting of a circuit, active after the first dropout line, which generated a CW signal corresponding to mid-grey in terms of off-tape frequency. The resulting neutral tone inserted in the picture was subjectively less noticeable than the dropout it replaced.

Later designs perform better by recirculating the last 'good' TV line around the delay line and reading it out continuously for the duration of the dropout. This gives better reproduction than the grey oscillator, and a recirculating DOC, based on an IC, is shown in *Figure 14.13*. The replay FM signal enters the AN316 chip at pin 2 and emerges, after amplification, at pin 3. After passing through a switched filter (colour-mono bandwidth switching by transistor X3) the signal re-enters the chip at pin 5 and passes through pole A of the dropout switch to emerge at pin 9, en route for the FM demodulator. A second output of the amplified RF signal appears at pin 4, routed through the limiter between pins 11 and 12, then into the dropout detector on pin 14. The dropout switching signal appearing at pin 13 is amplified in the source-follower X8 and reapplied to pin 10 of the chip, whence it operates the changeover switch to select pole B. Here the delayed signal is

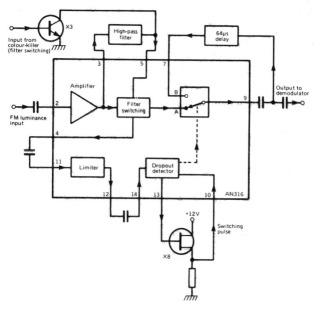

Figure 14.13. A recirculating dropout compensator designed round an AN316 chip (JVC-Ferguson)

available, and it is routed not only to the demodulator via IC pin 9, but back to the delay line input for recycling. This circuit requires only one demodulator.

Demodulation

Several methods of FM demodulation are available, and such configurations as the ratio detector and Foster-Seeley discriminator have been used for many years for FM sound recovery in radio and TV receivers. More up-to-date designs

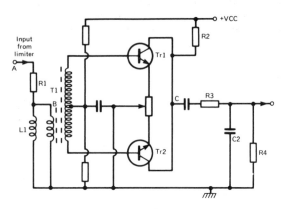

Figure 14.14. Pulse-counting FM demodulator

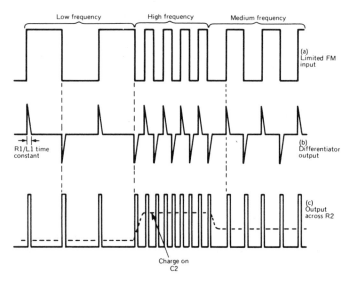

Figure 14.15. Waveforms for the demodulator of *Figure 14.14*. The frequency of the incoming squarewave (a) is effectively doubled at (c)

use a PLL (phase locked loop) or quadrature demodulator for better performance, but the low modulation index used in the VTR system (operating frequency 3–5 MHz, modulating bandwidth 0–3 MHz) requires a different technique. Two circuit configurations are common in home machines, the delay-line detector and the pulse-counting demodulator.

A circuit for the latter appears in *Figure 14.14*. The limited FM replay signal, represented by waveform (a) in *Figure 14.15*, is passed into the circuit at point A. R1 and L1 together form a differentiator, so that only spikes appear at the coupling transformer T1. The timing of the spikes depends on the FM frequency, but the duration of each is the same, set by the time-constant of R1 and L1. This is shown in *Figure 14.15b*. The secondary windings of T1 feed opposite-polarity spikes to the bases of detector transistors Tr1 and Tr2; Tr1 conducts on positive spikes and Tr2 on negative spikes, so that the frequency is effectively doubled at their common collector load resistor R2, giving rise to the waveform in *Figure 14.15c*. This frequency doubling, a form of full-wave rectification, greatly assists with filtering after the detector. All that remains, then, is to 'count' the pulses, and this is simply done by an integrator consisting of an RC or LC network, R3 and C2 in *Figure 14.14*. The time constant is arranged so that C2 is charged on each pulse and discharges through R4 between pulses – thus its average charge is proportional to pulse rate as shown by the dotted line in *Figure 14.15c*. After further filtering and de-emphasis a facsimile of the original waveform is obtained, now at baseband.

The demodulator circuit is carefully designed to offer linear detection throughout the deviation range, and the following filters effectively reject the residual (doubled) carrier ripple. The luminance signal we have now recreated consists of composite video (i.e. with sync pulses present) with a bandwidth of about 3 MHz.

Crispening

The 3 MHz capability of a standard VTR means that replayed pictures will lack the sharpness and definition of an off-air

269

picture; and fine detail, for example the frequency gratings of a test pattern, will not be reproduced. Several other factors are present, though, and these are worth examining. Most video recording involves colour programmes, and colour receivers incorporate notch filters in their luminance amplifiers which impair definition above 4 MHz. At about the same frequency physical limitations of the colour display tube come into play. The pitch of the phosphor-dot matrix on the tube faceplate is such that a picture element (*pixel*) smaller than the area of one RGB dot-group cannot be reproduced, and this is a limiting factor for luminance definition, especially in smaller colour screen sizes. Very little of the content of an average picture consists of fine repetitive detail, and the subjective effect of limited HF response is a blurring of sharp vertical edges.

The crispening circuit goes some way to compensate for HF roll-off by artificially 'sharpening up' vertical edges in the picture. The technique has been used in broadcast and CCTV for many years to compensate for the finite scanning spot size in cameras and FSS systems. In such applications it is known as aperture correction because it has the effect of limiting the scanning spot size to that of a smaller 'aperture' inserted into the tube, and enhancing definition thereby.

The effect of poor HF response is to slow down the rise and fall times of the luminance waveform so that a sudden transition from white to black (or vice-versa) is reproduced as at waveform (a) in *Figure 14.16*. There are many types of crispening circuit, and a block diagram of an easy-to-understand one appears in *Figure 14.17*. The luminance step waveform (a) is fed into a short delay line to emerge slightly later at the centre tap, waveform (b). Waveform (a) is also differentiated, producing a negative-going spike, waveform (c). After inversion this spike appears in positive-going form, waveform (d). After a further slight delay waveform (e) appears at the end of the delay line and is differentiated to a spike, waveform (f). Thus we have three carefully timed waveforms (b), (d) and (f) passing into the add matrix. Its output is shown as waveform (g), showing the *pre-shoot* effect of waveform (d) and *overshoot* effect of waveform (f).

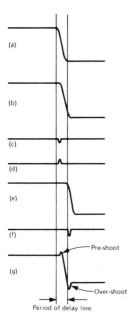

(a)

(b)

(c)

(d)

(e)

(f)

(g)

Pre-shoot

Over-shoot

Period of delay line

Figure 14.16. Waveforms for the crispening circuit of *Figure 14.17*. Their derivation is explained in the text

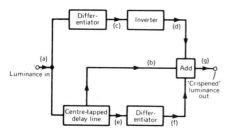

Figure 14.17. One form of crispening, or aperture-correction circuit

The slightly delayed signal now appears to have sharper transitions, and the process works on positive-going steps too. The delay time, represented by the two vertical lines in *Figure 14.16*, is carefully arranged along with the differentiator time-constants to give optimum correction at a chosen fre-

271

quency – study of *Figure 14.16* shows that above this optimum frequency the crispening circuit actually reduces definition! The degree of pre-shoot and overshoot is usually adjustable by a pre-set control labelled *crispening* or *aperture*.

Luminance delay

The luminance signal processing is now almost complete. Before the playback luminance signal is mixed with the off-tape chrominance signal, however, it is passed through a short delay line. This component serves the same purpose as the luminance delay line in a colour TV, and is necessary because of the differing bandwidths of the luminance and chrominance signal paths. The chrominance signal, being of relatively narrow bandwidth, has a slower rise time than the accompanying luminance signal, and if uncorrected this would have the effect of printing the chrominance component slightly to the right of the luminance image on the TV screen. The delay line, then, slightly retards the timing of the luminance signal to bring the two into step, ensuring correct *registration*.

After amplification, chroma addition, and conversion to low impedance, the vision signal is ready to pass out of the VTR. It is applied to the 'video out' socket in 1 V pk-pk form, and also to the RF modulator (along with the sound signal) where it is impressed on to a UHF or VHF carrier for injection into the aerial socket of a TV receiver.

S-VHS

The 'Super' variant of VHS format overcomes the definition limitations of conventional formats. Here the FM carrier frequency deviates between 5.4 and 7.0 MHz, a 60 per cent expansion. The luminance sidebands are correspondingly wide; the entire luminance FM 'package' occupies a bandwidth of over 8MHz, compared with 5 MHz in the original plan. S-VHS cassettes have similar running time and appearance to standard types, but contain cobalt-oxide coated tape and can reproduce pictures with 400-line resolution.

15

Signal processing: chrominance

To understand the colour circuits of a VTR, it is essential to have a working knowledge of the way in which chrominance signals are encoded in ordinary broadcast transmissions. Full details are given in Chapters 3 and 4, and revision of those sections of the book is recommended to readers as a basis for study of the colour section of a VTR. One of the advantages of home VTRs is that at no point is the colour signal demodulated – it remains in subcarrier form throughout the system, giving little opportunity for phase (thus hue) errors. A short recap of the PAL system follows.

PAL colour recap

A TV system requires three simultaneous streams of information to produce a full colour picture; these are the R, G and B (red, green and blue) video signals. For a monochrome signal, such as a black-and-white film, all three are identical and a single signal channel (i.e. the single electron gun and signal path of a monochrome TV set) is all that is required to convey the information. For colour transmissions it is impractical and wasteful to use three separate channels for the RGB signals, and so a method was devised to encode the colouring signals in such a way that they 'ride piggyback', as it were, on the basic luminance signal and the whole can be accommodated within a single 8 MHz side transmission channel.

At the sending end the luminance signal is transmitted in a form represented by the waveforms of *Figure 13.14*. This satisfies a monochrome receiver, and conveys the basic picture – and fine detail – to a colour set. Instead of sending RGB (known as *primary colour-*) signals, *colour-difference signals* are produced and these represent, for each primary colour, the difference between the luminance signal and the desired colour at any instant. It is not necessary to send all three colour-difference signals; two will do, because the third can be recreated at the receiver by a matrix fed by the two colour-difference signals we do have, plus the luminance signal. The two colour-difference signals chosen for transmission are those for red and blue, represented by R-Y and B-Y respectively, and each of these amplitude-modulates a *subcarrier* signal at approximately 4.433 MHz. These two amplitude-modulated subcarrier signals, although at the same frequency, are 90° apart in phase, which means that one lags the other by one-quarter of a cycle; they are said to be in *quadrature*. Because *balanced* chrominance modulators are used at the transmitter the subcarrier itself is suppressed, and only the sidebands, or modulation products, appear at the modulator output. The system is called *suppressed carrier modulation*, and depends entirely on the sidebands of the signal to carry the information. Elimination of the carrier (or subcarrier in a colour TV system) has an important advantage. Lying as they do within the luminance passband, any chrominance signals, however coded, will upset the luminance reproduction by causing spurious patterning in the form of dots and lines, and this will happen on monochrome and colour receivers. Careful selection of the subcarrier frequency (4.43361875 MHz, locked to line frequency) minimises the chrominance/luminance crosstalk, and this is further reduced by the suppressed subcarrier technique which ensures that dot-patterns occur only in coloured areas of the picture, and are noticeable only on highly-saturated colours.

Phases and vectors

The subject of phase relationships and vector diagrams is not a widely-understood one. We have already touched on the subject of phase in describing the colour-encoding process, and it will crop up many times in our study of VTR chrominance and servo systems. The word 'phase' describes the timing relationship between two waveforms, usually – but not necessarily – at the same frequency. In *Figure 15.1a* is shown a sinewave which we will call the *reference*. b shows

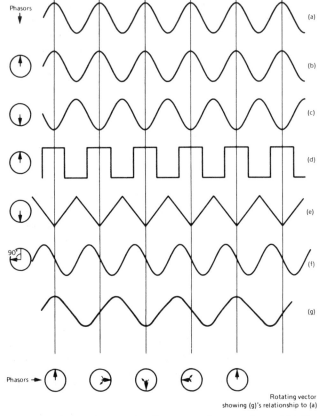

Figure 15.1. Phase relationships between various waveforms. (a) is the reference and the 'clocks' on the left show vector relationships. Waveform (g) is at a frequency lower than that of the reference (a), giving rise to the clockwise vector rotation shown at the bottom

an identical sinewave with the same timing as *a* and this is said to be 'in phase' with the reference, as shown by the vector diagram alongside. One complete trip round the vector diagram (360°) represents one cycle of the reference signal, so that waveform *c*, being half a cycle behind the reference signal, is phase-retarded by 180°, or half a circle; it is said to be *anti-phase*. It will be noticed that waveform *c* is an 'upside-down' version of the reference *a*, so that inverting a waveform *if it is symmetrical about the zero line over one cycle* (i.e. a sine-wave or square-wave) will give the same effect as a phase change of 180°. The waveforms involved do not have to be sinusoidal of course; *d* shows a square-wave in phase with the reference and *e* a triangular wave in antiphase, or at 180°.

When two waveforms have a phase relationship of 90° or 270° they are said to be in *quadrature* and their vectors are at right-angles to each other; this quadrature relationship is a useful one, because it means that when the reference waveform is at its peak, its quadrature-companion is passing through zero, so that by 'sampling' at the correct instant we can separate the information carried by each of two mixed signals in quadrature – without crosstalk. This, of course, is the principle used in the synchronous demodulators of colour decoders in TV sets. Waveform *f* in *Figure 15.1* is in quadrature with the reference *a*.

Rotating vectors

If we introduce a second signal at a frequency slightly lower than that of our reference, the situation will be as in *Figure 15.1g*. Here the waveforms start in phase, but the lower frequency (g) will fall behind the reference as time goes on, so its phase lag will increase with time, giving rise to the clockwise-rotating vector shown. After a period, the second signal will be exactly 1 cycle behind the reference, and the two will be momentarily in phase again, with the vector having completed one full clockwise circle. For a frequency higher than the reference signal, it can be seen that the vector will rotate anticlockwise; in both cases, the speed of rotation of the vector will be proportional to the difference in frequency between the reference and the sample.

If the frequency of the sample is constantly changing with respect to the reference, but by less than one cycle, the result is a 'phase jitter', represented by a constantly-changing vector angle, but confined to less than 360° (one full turn). This is what happens, for instance, in a colour TV system, where the reference signal is a stable subcarrier source, and the 'sample' is the transmitted chroma signal.

Phase-locked loop (PLL)

A common 'building block' in VTR systems is the phase-locked loop. It takes the form of an oscillator and phase detector, and is designed to produce an output signal with a fixed phase relationship to a reference signal; a familiar example is the flywheel line timebase circuit of a TV receiver.

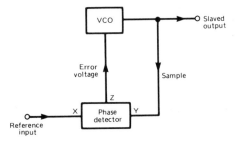

Figure 15.2. Basic operation of a phase-lock-loop

A simple example of a PLL is shown in *Figure 15.2*. The process starts with a phase detector, or discriminator, consisting of a device with two inputs, X and Y. When the timing of input pulse X is earlier than that of input pulse Y, the detector will produce a positive DC voltage at its output Z. If, alternatively, pulse Y occurs before pulse X, output Z will move negatively. When both input pulses are coincident in time (i.e. *in phase*) output Z is zero. Thus the output Z of the phase detector

278

indicates by its polarity the direction of phase error, and by its amplitude the amount of phase error. Not surprisingly, this is called the error voltage!

The second half of the PLL is a voltage-controlled oscillator (VCO). This is an oscillator whose output frequency – and phase – can be controlled by a DC voltage. It may consist of a crystal oscillator (as in the reference generator of a colour TV decoder), or an LC or flip-flop oscillator. Its output forms one input to the phase detector, and its frequency-controlling input is the error voltage from the phase detector. Thus is the 'loop' set up, and whenever the oscillator and input signals drift apart in timing, the discriminator generates an error signal to pull the oscillator back into phase lock with the control pulse, and again achieve zero phase difference between its X and Y inputs. The oscillator is then said to be 'slaved' to the reference, or 'master', signal.

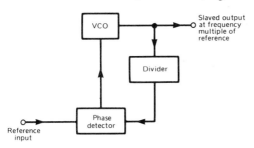

Figure 15.3. Phase-lock loop with divider. The output becomes locked to a multiple of the reference frequency

The oscillator of a phase-locked loop does not necessarily run at the same frequency as the control pulse; with a suitable divider in the feedback path from the oscillator to the phase detector, an output locked to a multiple of the 'master' frequency can be obtained, as shown in *Figure 15.3*.

Quadrature modulation

Addition of the outputs of the R-Y and B-Y modulators at the transmitter renders a single chrominance signal (consisting of just the sidebands of a 4.43 MHz sinewave) which conveys the hue of the colour by its phase, and the saturation of the colour by its amplitude. This chrominance signal is mixed with the luminance signal before transmission, and being based on a frequency of 4.43 MHz it is easily filtered out for decoding within the colour receiver. *Figure 15.4* shows the spectrum of the colour-encoded video signal.

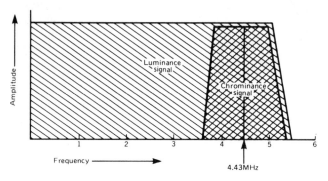

Figure 15.4. Spectrum of demodulated broadcast video signal

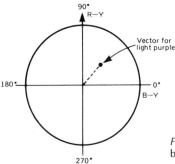

Figure 15.5. Vector diagram of broadcast colour encoding

A vector diagram, *Figure 15.5*, puts the colour subcarrier system into perspective. The dotted line represents the phase of the subcarrier signal, and this can extend to any amplitude (length change) to describe saturation; and point in any direction (angle change) to describe hue. Thus the

280

short vector pointing NE illustrated represents a light purple colour; a long vector pointing SW would indicate a bright green colour.

Colour burst

At the receiver we need to recreate the missing subcarrier signal in order to demodulate the chrominance signals. The recreated subcarrier must be in phase with that at the transmitter – phase, as we have seen, means relative timing, and to measure the phase or timing of the colour subcarrier signal, we need a reference against which to do it. A timing reference is transmitted once per line, then, and it takes the form of a 'sample' of ten cycles of plain subcarrier sitting on the back porch of the video signal, illustrated in *Figure 15.6*, and known as the *colour burst*. This is gated out in the TV's decoder and used to phase-lock a crystal oscillator, thus providing the necessary timing reference.

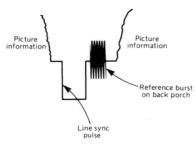

Figure 15.6. Position of the reference burst in the line waveform

An essential feature of the PAL system is the effective inversion of the R-Y subcarrier signal on alternate lines. This is done to eliminate hue errors due to phase distortion in the signal path, and facilitates cancellation of such errors in a delay-line matrix circuit. Thus on TV line 1 the R-Y axis is OB in *Figure 15.7*, whereas on line 2 the R-Y axis is OD. The B-Y axis is OA on all lines, so that magenta (a shade of purple containing equal proportions of red and blue) would appear as vector OX on line 1, OY on line 2, OX on line 3 and so on.

281

When its usefulness has been realised (after the delay-line), this phase alternation has to be cancelled out by an inverting switch working at line rate. This is the PAL switch, and to keep it in synchronisation a further signal, called *ident* (short for identification), is sent. This is achieved by advancing and retarding the phase of the burst signal itself by 45° per line – the *swinging burst*. The burst phase alternations, then, are used in the TV's decoder to identify the PAL line (the one

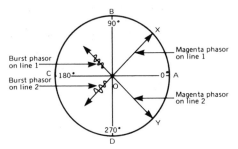

Figure 15.7. Colour and burst axes for PAL encoding. The burst and V chroma signals are 'mirrored' about the axis AC on a line-by-line axis

on which the R-Y carrier is inverted), while the reference oscillator locks to the *average* burst phase. *Figure 15.7* indicates the relationship between burst phase and R-Y subcarrier phase.

Chrominance bandwidth

In transmission, the bandwidth of each colour-difference signal is limited to 1.2 MHz; this results in a rough, low-definition colour image which, however, is very acceptable when overlaid on the detailed luminance picture. This is because the human eye cannot perceive fine detail in colour, and the brain is fooled into believing that a full-definition, full-colour picture is present. We shall see that further liberties are taken with chrominance bandwidth in domestic VTRs, but still with subjectively acceptable results.

282

Colour signals in the VTR

If we had a perfect video tape recorder there would be no reason why we could not take the composite video signal, chrominance and all, and record it on the tape as an FM signal, demodulate it during replay and recreate the original video signal as described for luminance in Chapter 14. As we've already seen, a home VTR is far from perfect in its performance, and the direct recording idea will not work for two good reasons.

The bandwidth limitations of our VTRs have already been described, and with an upper cut-off frequency around 3.5 MHz, the colour signals at 4.43 MHz are 'left out in the cold' somewhat! Even in S-VHS and Beta-ED formats, whose signal passband approaches 5 MHz, it is still not possible to directly record an encoded colour signal. Because the VTR relies on mechanical transport, perfectly smooth motion cannot be achieved. Inevitable imperfections in bearing surfaces and friction drives, the elasticity of the tape and minute changes in tape/head contact all combine to impart a degree of timing jitter to the replayed signal. In good audio tape systems this jitter is imperceptible, and for luminance replay its effects can be overcome by the use of timebase correctors in professional VTRs, or fast-acting TV line-synchronising systems with domestic VTRs. The latter class of machine, in good condition, may be expected to have a timing jitter of about 20 μs over one 20 ms field period. Let's examine the effect of this on the chrominance signal.

We have seen that the hue of a reproduced colour picture depends entirely on the phase of the subcarrier signal. 'Phase' really means relative timing, so if the timing of the subcarrier signal relative to the burst is upset, hue errors will appear. One cycle of subcarrier occupies 226 ns, and a timing error of this order will take the vector right round the 'clock' of *Figure 15.5*, passing through every other hue on the way! A timing error of 100 ns (one ten-millionth of a second) will turn a blue sky into a lime-green one, and the jitter present in a mechanical reproduction system will make nonsense of the colours in the picture. For acceptable results, subcarrier phase errors need to be held within 5° or so, representing a permissible maximum timing jitter of ±3 ns as shown in *Figure 15.8*. The physical causes of jitter cannot be eliminated

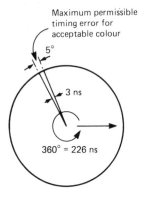

Figure 15.8. Hue is dependent on phase angle, and only a very small phase change is permissible in a VTR system

so an electronic method of subcarrier phase correction is necessary.

Colour under

Let's first examine the problem posed by the high colour subcarrier frequency. Several solutions are possible, including decoding the chrominance signal to baseband (i.e. primary colour or colour-difference signals) before committing them to tape, then recoding them to PAL standard during the replay process. This would require a lot of circuitry, and be vulnerable to hue and saturation errors. A better solution is to keep the PAL-encoded signal intact, bursts, phase modulation and all, and merely convert it to a suitable low frequency for recording. During playback the composite chrominance signal can be re-converted to the normal 4.43 MHz carrier for recombination with the luminance component into the standard form of colour video signal, compatible with an ordinary TV set.

Space has to be found for the colour signal in the restricted tape frequency spectrum, however, and it is made available

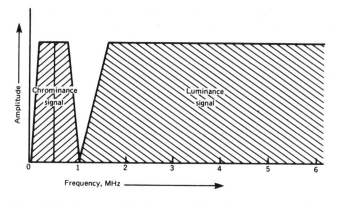

Figure 15.9. Spectrum of luminance and chrominance signals on the tape. The luminance signal is now in FM form, and the chroma signal still at baseband, but with limited bandwidth

below the luminance sideband, where a gap is purposely left in the 0–1 MHz region. As explained in Chapter 14, this area is kept clear by tailoring the luminance record passband and the FM modulator deviation to avoid luminance sidebands falling into it during colour recording. The colour signal, then, is down-converted and bandwidth-limited so that it occupies a frequency band of roughly 0–1 MHz. It is a double-sideband signal, so that the carrier frequency needs to be centred on about 500 kHz and the sidebands limited to 500 kHz or so. The spectrum of the colour-under signal is shown in *Figure 15.9*; compare this with *Figure 15.4*.

Down-conversion

The process for down-converting the chrominance signal is shown in *Figure 15.10*. The double-sideband signal is first separated from luminance information in a high-pass filter, then its bandwidth is limited to around 1 MHz by a bandpass filter. This curtailment of the sidebands of the modulated signal has the effect of restricting the chroma signal detail to around 500 kHz, less than half that of the broadcast signal.

The chroma signal now passes into a mixer where it beats with a stable locally-generated CW signal at around 5 MHz. This is the familiar heterodyne effect, and the mixer output contains components at two frequencies, those of the sum and difference of the two input signals. A low-pass filter selects the required colour-under signal, and rejects the spurious HF product.

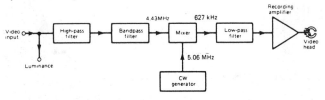

Figure 15.10.   The basic colour-under system. A heterodyne technique is used to convert the chroma signal to a new, low frequency

The colour-under signal contains all the information that was present in the 4.43 MHz chroma signal except the finer colour detail; thus chroma amplitude and phase modulation are preserved, as well as the phase and ident characteristics of the burst, albeit based on a much lower subcarrier frequency. *Figure 15.11(a)* shows an oscillogram of the chrominance component of a colour-bar signal in 4.43 MHz form as broadcast, while in (b) appears the same signal after down-conversion to a base frequency of around 500 kHz. Looking at the coarse structure of the waveform at *(b)* it is hard to believe that all information is preserved intact, especially the crucial burst signal, which appears as little more than two cycles of oscillation – it is so, however!

Up to now we have been quoting subcarrier frequencies in round numbers. Format specifications quote the colour-under frequency very precisely, and taking VHS as an example, the local CW signal is at a frequency of 5.060571 MHz, which when beat against the 4.433619 MHz broadcast subcarrier provides a colour-under centre frequency of $5.060571 - 4.433619 = 0.626952$ MHz, or 626.952 kHz. Due to broadcast subcarrier modulation, sidebands will extend for roughly 500 kHz on each side of this carrier.

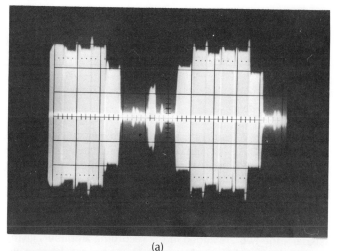

(a)

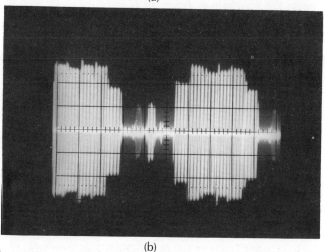

(b)

Figure 15.11. (a) Oscillogram of encoded colour-bars as broadcast, based on a frequency of 4.43 MHz. (b) Down-converted colour-bars, now at a basic frequency about one-eighth of that in (a). The 'ghost' burst to the left of the real one is a spurious effect of the oscilloscope from which the photograph was taken, and is not actually present in the colour-under waveform

For reasons which will become clear later, the local CW signal at 5.06 MHz needs to be locked (i.e. have a fixed phase relationship) to the line syncs of the recorded signal and this is achieved by a PLL (phase-locked loop).

Chroma recording

The down-converted colour-signal based (for standard VHS format) on 627 kHz, and containing amplitude and phase modulation, is recorded directly on to the tape. Its amplitude is carefully controlled to achieve maximum modulation of the tape without non-linearity due to magnetic saturation. In this respect it is similar to audio tape recording, with the necessary HF bias being provided by the FM luminance record signal – the two are added in the recording amplifier, giving rise to the waveform of *Figure 15.12*, where can be seen the recording-head signal for a colour-bar signal. The chrominance signal is superimposed on the constant-level FM

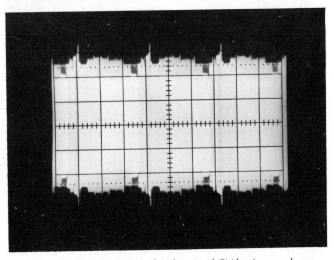

Figure 15.12. The composite colour bar signal (FM luminance plus down-converted chroma) as applied to the recording heads

luminance carrier, and this composite waveform is fed to the heads for recording on the tape.

Chrominance replay

During playback the chrominance signal appears at the replay amplifier in the colour-under form described above, and with a waveform like that in *Figure 15.12*. Impressed on to it are the timing variations, or jitter, introduced by the mechanical record and replay transport system and these have to be eliminated. First, though, the chroma signal is separated from the luminance FM signal in a low-pass filter. Up-conversion to correct subcarrier frequency (4.43 MHz) must also be carried out, and we'll take this first.

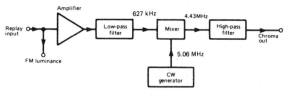

Figure 15.13. Basic chroma replay process of up-conversion

The 627 kHz is heterodyned with a local 5.06 MHz CW signal to produce a beat frequency of 4.43 MHz, selected by a suitable bandpass filter. The idea is shown in *Figure 15.13*. Dropout compensation and crispening are not relevant to the relatively coarse and 'woolly' colour replay signal, and the recreated 4.43 MHz chroma signal is added to the luminance component and passed out of the machine.

De-jittering

As we have seen, the up-conversion process during replay involves a heterodyne system. If we can ensure that the 'local oscillator' is modulated by just the same timing errors as the off-tape chroma signal, the jitter on the replayed waveform will be cancelled out. The heterodyne process works by producing a *difference frequency* between two inputs (in this case the local 5.06 MHz CW signal and the off-tape chroma signal at 627 kHz). If both inputs contain the same jitter component they will move together in frequency and phase, so that the difference between them will be constant. This 'constant difference' is in fact the 4.43 MHz output signal, which will thus be unaffected by timing errors. On replay, then, the 5.06 MHz CW signal is known as the *jittering reference*. Its derivation will now be described.

The luminance signal recorded on the tape contains line sync pulses, and when these are recovered from the luminance FM demodulator they will contain the same timing errors, or jitter, as the chrominance signal, having been recorded and played back under the same circumstances and at the same time. If we separate them from the luminance part, they can be used to generate a suitable jittering reference. *Figure 15.14* shows how. Here we have a stable crystal oscillator running at 4.435 MHz as one input to a mixer, and a

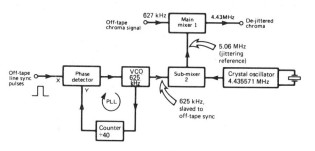

Figure 15.14. A jittering reference generator using off-tape line sync and a phase-lock loop

625 kHz signal as the other. The 625 kHz signal is generated by a VCO (voltage-controlled oscillator). Now 625 kHz is exactly 40 times line frequency, so if we divide the oscillator output by 40 in a *counter* we have a line-rate signal suitable for comparison with off-tape line sync in a phase detector. This is another example of a PLL and its effect is to

290

impress the jitter signal on to the VCO output. This jittering 625 kHz signal when beat against the stable 4.43 MHz crystal output will produce a jittering 5.06 MHz reference as required.

Let's take an example to illustrate the action of the circuit. Assume that a timing error has momentarily increased the off-tape chroma carrier frequency by 40 Hz, bringing it up to 626.992 kHz. Because colour-under frequency is 40 times line rate, the off-tape line sync pulse frequency will simultaneously rise by 1 Hz to 15.626 kHz, and this will appear at point X on *Figure 15.14*. The phase-lock loop will push up the VCO frequency to rebalance its phase detector and achieve 15.626 kHz at its Y input, and VCO output will rise to 625.040 kHz. The sub-mixer 2 will beat this against the steady 4.435571 MHz crystal frequency to produce an output of 5.060611 MHz. This passes into the main mixer 1 where it beats with the off-tape chroma carrier at 626.992 kHz to produce a final output frequency of 5.060611-0.626992 = 4.433619MHz as was present with no timing error. Although this example describes a *frequency* error, phase errors (timing errors

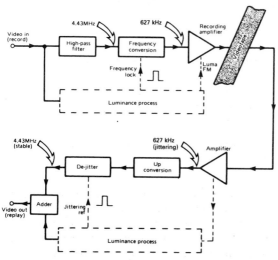

Figure 15.15. Overall view of the colour-under system

291

of less than one down-converted subcarrier cycle) are dealt with in the same way.

It is now plain why the colour-under signal is phase-locked to off-air line sync during record – the sync and colour signals are being recorded simultaneously, so that jitter cancellation works for both record and playback processes, the key in each case being the closely accompanying line sync pulse. The same PLL and 40-times divider is used in record and playback. The block diagram of *Figure 15.15* shows the basic processes in chrominance recording, playback and jitter-compensation in a domestic VTR.

There are several other methods of deriving a jittering reference, all involving an off-tape signal of one kind or another. In the *burst-lock* system the colour-burst signal itself is used, being gated out of the replay chrominance signal and used to slave a local oscillator. Another system uses a pilot tone, laid down on the tape during record, to achieve the same end. All current home formats, however, use the line-sync reference system described above, in conjunction with a 'tighter' jittering reference derived from the off-tape chroma sub-carrier signal itself – more details are given later.

ACC and colour-killers

Colour TV sets incorporate an ACC (automatic colour control) system to prevent tuning and propagation errors upsetting the luminance/chrominance ratio. This is necessary in a VTR for the same reasons, so that during record the ACC circuit monitors the amplitude of the colour-burst signal (transmitted at a constant level) and adjusts the gain of the chrominance record amplifier to maintain it at a constant level; by these means the chrominance signal itself is held at correct amplitude. To prevent the recording of *confetti* or unlocked colours during monochrome transmissions or when the record down-conversion PLL is unlocked, a record colour-killer is provided, and this is sometimes linked to a rear-mounted switch so that killer action can be invoked manually for a full-bandwidth monochrome recording of any signal which contains colour information.

On replay, too, these functions are required. The off-tape chrominance signal is vulnerable to level fluctuations and these are ironed out by the ACC loop. The replay colour-killer (which again can be manually overridden in some machines) acts to prevent spurious colouring of replay pictures from monochrome videotapes, and shuts down the chrominance channel in the event of malfunction of the replay up-conversion processes.

Crosstalk compensation

In Chapter 13 we saw that early machines used a guard band between recorded tracks on the videotape to prevent the possibility of crosstalk between adjacent tracks. Later formats eliminate this guardband and achieve greater information density on the tape by laying down tracks with no intervening safety gap. To prevent crosstalk between tracks each head is given an azimuth offset; this ensures that readout from adjacent tracks is at a very low level and does not interfere

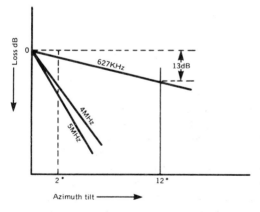

Figure 15.16. The effects of azimuth tilting at different frequencies. LF signals have low immunity to crosstalk

with reproduction. As *Figure 15.16* shows, however, the azimuth-offset idea is less effective at low frequencies, and the 0–1 MHz region occupied by the chroma signal will suffer unacceptable crosstalk between tracks unless steps are taken to remove the interfering signal.

We cannot prevent inter-track crosstalk taking place at colour-under frequencies, so methods of electronic cancellation of the unwanted signal have been devised. Each format uses the same basic idea but differs in the way that it is carried out. Let's look at a typical cancellation system, that used in the VHS format, to explain the principles involved.

The interlaced scanning system employed in broadcast TV calls for a half-line offset between the starting points of odd

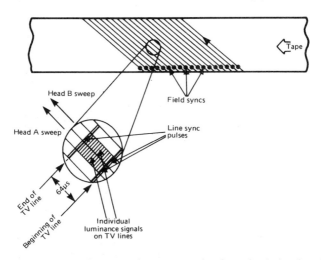

Figure 15.17. The tape tracks are arranged so that individual TV lines lie physically adjacent on the tape

and even fields, as explained in Chapter 1. If the video tracks were laid down on the tape in this form, there would be a half-line offset betwen the recorded TV lines, just like that shown on the TV screen in *Figure 13.13*. However, it is not that simple, as the video tracks are laid on the tape at a slant

294

angle, and the tape moves horizontally past the drum. The combined effects of these result in the individual lines of each TV field being laid exactly alongside each other, as shown in *Figure 15.17*. This ensures that when crosstalk occurs the interfering signal is coming from the 'same' line on the partnering field, and has the *same timing*, within that line, as the wanted signal. Thus there will be a fixed phase relationship between wanted and crosstalk signals, bearing in mind that the colour-under signal is based on a frequency (for the VHS system) of 626.9 kHz.

During record, head A is fed direct by the colour-under signal, and the PAL characteristic of this signal will result in the burst phases being laid down on the tape according to (a) in the chart of *Figure 15.18*. Thus TV line *n* has its burst at 135°,

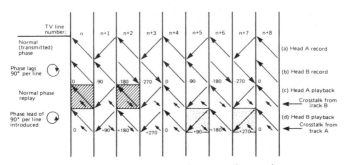

Figure 15.18. Burst phasor diagram for the VHS colour-under system. Lines (c) and (d) can be seen to emerge with correct phasors as broadcast

line *n*+1 at 225°, line *n*+3 135° and so on. The chroma signal fed to head B is *delayed* by 90° on each successive TV line as shown in line (b) of the chart. Thus line *n* is 'normal', line *n*+1 has a 90° clockwise shift, line *n*+2 is rotated 180° clockwise, line *n*+3 270° clockwise, line *n*+4 normal, and so on throughout the field period. During replay, head A will pick up a large signal from its own track and a smaller crosstalk signal from the adjacent tracks laid down by head B, with burst phases as shown in line (c) of the chart. To recover a 'normal' signal from head B during replay, we have to

295

advance the phase of its output by 90° per TV line, to compensate for the 90° retard per line introduced during recording. Thus head B's corrected output will be line (b) on the chart, but phase advanced by 90° per TV line, as shown on chart line (d). The small arrows here show the crosstalk signals picked up from track A, *having undergone the same 90° per line phase advance*. Study of playback lines (c) and (d) of the chart will show that if we compare *any* playback TV line with its next-but-one neighbour, the main signal vectors add (i.e. they are pointing in the same direction), whereas the crosstalk vectors (small arrows) cancel. This is illustrated by the hatched boxes for head A playback (TV lines *n* and *n*+2) and the dotted boxes for head B playback (TV lines *n*+5 and *n*+7).

The reason for the carefully-contrived head B recording sequence is now clear. To eliminate the crosstalk signal it is only necessary to *add* each TV line to its next-but-one neighbour, which can be brought into time coincidence by a two-line (128 μs) delay, then added in a suitable matrix, as illustrated in *Figure 15.19*. The process is very similar to that which takes place in the familiar delay line circuit of a colour TV receiver, and the glass delay line is similar in construction (but with twice the path-length) to its TV counterpart.

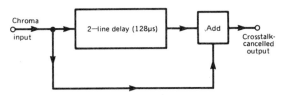

Figure 15.19. Delay line matrix for chroma crosstalk cancellation

We have described the crosstalk-compensation process in terms of burst phasors, whereas our real concern is eliminating crosstalk between the chrominance *signals* on adjacent tracks. Since the chroma signal is defined (in its encoded form) in amplitude and phase, the crosstalk compensation system works equally well on the subcarrier signal as on the burst itself.

296

Before we go on to describe the electronics in the chroma phase-shifting process, we must recap the relevant parameters of the VHS format. The colour-under frequency is 626.952 kHz, produced in a balanced modulator (mixer) by beating the incoming 4.433619 MHz colour signal with a locally generated 5.060571 MHz CW signal containing the necessary phase-shifting information for recording head B.

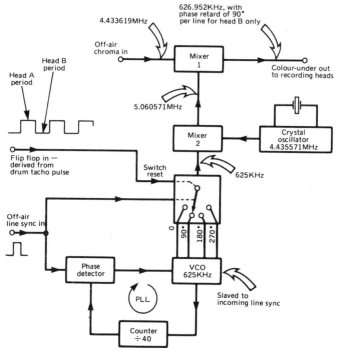

Figure 15.20. The VHS colour recording system to give the tape track pattern of *Figure 15.18.*. Operation is explained in the text

Figure 15.20 shows the system in block diagram form. This is similar to the colour-under system we met earlier in this chapter, with a main mixer 1 producing the colour-under signal and a sub-mixer 2 acting to produce the 'local' signal.

For VHS, mixer 2 is fed from the output of a stable free-running crystal oscillator at 4.435571 MHz; and a 625 kHz signal (40×line frequency) generated by a VCO slaved to incoming line sync. On its way to mixer 2, the 625 kHz signal passes through a phase shifter (part of a purpose-designed IC) which *advances* its phase by 90° per line for one field (that recorded by head B) but is reset to zero phase shift for the next field (head A's province). The 'reset' signal comes from a tachogenerator pulse which is derived from a magnet on the head drum, and which indicates that head A is writing. This 90° line-by-line phase advance, operating on each B track only, passes intact through additive sub-mixer 2, and is turned into a phase *retard* at subtractive main mixer 1 for application to recording head B.

To sum up, then, we have going on to the tape a down-converted subcarrier signal based on 626.9 kHz, with side-bands extending down to about 100 kHz and up to about 1.1 MHz. This subcarrier is identical to the encoded broadcast colour signal, but based on the new, lower frequency. Head A records odd-numbered tape tracks without modification, whereas head B records the subcarrier signal on even-numbered tape tracks with a phase retard of 90° on each TV line. The recording signal for head B is modified in this way to facilitate crosstalk elimination in a 2-line delay matrix during replay.

VHS chroma replay

Figure 15.21 shows a simplified block diagram of the replay process. The off-tape 626.9 kHz chroma replay signal enters at the top left, and is mixed in main mixer 1 with the local 5.06 MHz signal to produce a difference signal of 4.43 MHz, selected by the bandpass filter. This forms the up-converted output signal. The local 5.06 MHz signal needs to have a phase-stepping characteristic to compensate for that intro-duced during record. Once again, sub-mixer 2 is used, and operates, as on record, to beat two locally produced signals – these being the 4.435571 MHz crystal oscillator output; and the 625 kHz (40×line frequency) signal from the phase-lock loop. The action of the phase shifter is the same as on record (phase advance of 90° per line for head B only) but because both mixers, 1 and 2, are now operating in *subtractive* mode the phase advance effect emerges unchanged and appears as a phase advance at the output of main mixer 1, cancelling the phase retard effect imparted during the record process.

The recovered and up-converted chroma signal now passes into a 2-line (128 μs) delay line matrix, where crosstalk cancellation takes place to render a 'clean' chrominance signal at 4.43 MHz, ready to be added to the luminance signal.

VHS de-jittering

The off-tape chroma signal at 626.9 kHz contains a timing error (jitter) component, and this has to be removed during replay. As in the 'basic' process described earlier in this chapter, this is achieved by using a jittering reference from off-tape line sync. The reference signal for the 625 kHz PLL at the bottom of *Figure 15.21*, then, comes from the replay luminance signal via a sync separator, and the 625 kHz input to sub-mixer 2 is thus slaved to off-tape sync, and contains a jitter reference as well as the 90° per line phase stepping effect for head B. The jittering reference passes through sub-mixer 2 and cancels the chroma jitter in main mixer 1.

A further de-jittering process is carried out in VHS machines, and this is shown in *Figure 15.22*, an enlargement of the previous diagram *Figure 15.21*. Here we see the additional circuit blocks on the right, and their operation will now be explained. A sample of the 4.43 MHz chroma output is taken and passed into a burst gate which extracts the colour burst signal. This burst is compared with the output of a stable 4.43 MHz crystal oscillator in phase detector 1, any jitter present giving rise to an error signal which 'steers' the frequency of the 4.435571 MHz crystal oscillator (now working as a voltage-controlled oscillator, VCO) in a direction to compensate for the jitter, and cancel it out. If the phase error becomes too great for phase detector 1 to cope with, phase

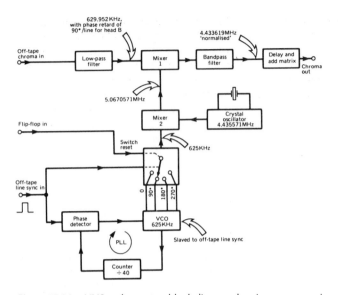

Figure 15.21. VHS replay system block diagram showing up-conversion, line-rate de-jittering and phase restoration

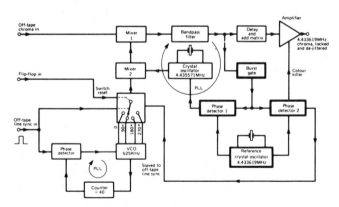

Figure 15.22. Expansion of *Figure 15.21* to show the second de-jittering loop, which works at subcarrier rate. Newly introduced blocks are shown in bold outline

detector 2 comes into play, sending a pulse to the phase-select circuit to shift the 625 kHz signal by 180°, thus pulling it back within the range of phase detector 1. The net result is that the 4.43 MHz chrominance output signal is slaved to the stable 4.43 MHz crystal oscillator reference.

A colour-killer output is taken from phase detector 2, and this ensures that replay chrominance is only enabled when the VTR's chrominance output is correctly locked to the local 4.43MHz crystal oscillator.

Betamax colour system

We have seen that the elimination of chroma crosstalk in the VHS system relies on the principle of laying down adjacent tracks on the tape so that individual TV lines lie alongside one another, and adjusting the phase of the recorded colour-under carrier so that over any 2-line period the main playback signals are in phase whereas the unwanted crosstalk signals are in antiphase. This facilitates the use of a 2-line delay and add-matrix to cancel the crosstalk component. The Betamax format uses the same principle and the same delay-line matrix. Its main difference from VHS is in the method of deriving the chrominance recording signal to achieve the two-line phasing effect.

In the Betamax system, the basic colour-under carrier is at 687.5 kHz, but each head is given a frequency offset from this figure – head A operates 1.953 kHz below the basic frequency, and head B 1.953 kHz above it. These figures have been carefully chosen, in conjunction with the line scanning rate of 15.625 kHz, to give the effect of a phase *retard* of 45° per line for head A, and a phase *advance* of 45° per line for head B. Thus the rotating vectors effect described on page 276 is set up, with the vector for head A rotating clockwise (with respect to line sync) and taking 8 TV lines per revolution, and the vector for head B rotating anticlockwise, and taking 8 TV lines per revolution. This effect is shown in lines (a) and (b) of the chart of *Figure 15.23*, wherein 'normal' (transmitted) burst phasors are shown in dotted outline for each head during

301

record, and the actual phases recorded on tape as full arrows. Thus for each head, the recorded subcarrier phase will return to normality – and agree with the transmitted PAL phase – every eight lines.

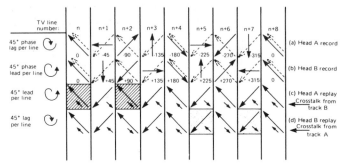

Figure 15.23. Burst phasor diagram for Betamax colour-under system, with transmitted burst phases in dotted outline. Although the record and playback phase shift sequence is different to that of the VHS and Video 8 system, the net result is just the same - compare with *Figures 15.18 and 15.29*

Now, because the recorded chrominance vectors for the heads are rotating in opposite directions at 45° per TV line, the vector for head B is effectively advancing by 90° per TV line with respect to that for head A, so that the phase angles laid down on the tape are in a four-line pattern. On replay, the phasors come out as shown in lines (c) and (d) of the chart of *Figure 15.23*. These can be seen to be very similar to lines (c) and (d) of the VHS phasor chart of *Figure 15.18*, and the same lines (*n* and *n+2*, then *n+5* and *n+7*) have been picked out to show the cancellation effect which takes place in the delay line matrix previously described. This is another way of saying that the Betamax format has the same (90° per TV line) chrominance phase offset between heads as the VHS system, but instead of holding one head's phase steady and rotating the other clockwise at 90°/line, the head phasors are rotated in opposite directions at 45°/line to give the same effect.

302

Pilot burst recording

A unique feature of the Betamax chrominance system is the insertion into the recorded colour signal of a pilot burst signal. In our study of chroma de-jittering techniques we saw that early domestic VTRs relied on the off-tape line sync pulse alone to generate a jittering reference during replay, and that the VHS system has an additional loop in which the off-tape swinging burst signal (an integral part of the encoded colour signal) is used to check and correct any remaining chrominance jitter. The Betamax system adds a constant-phase pilot

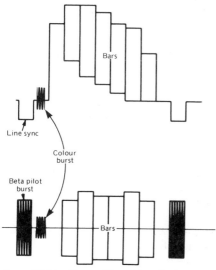

Figure 15.24. The position of the pilot burst in the Betamax colour-under system

burst to the chrominance signal during the recording process (it is inserted during line blanking, as shown in *Figure 15.24*) for use as a jittering reference on replay.

Betamax record system

The colour-under carrier for the Beta chrominance record system is produced in similar fashion to that for VHS, but instead of a phase-switcher in one head channel, we need what amounts to a two-tone generator to alternately produce the correct frequencies for heads A and B. A simplified block diagram of the system appears in *Figure 15.25*. Starting at top

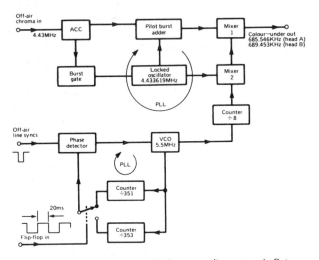

Figure 15.25. Block diagram of colour recording system in Betamax machines

left, the filtered off-air chroma signal passes through an ACC stage before entering the pilot-burst adder stage, where a sample of locally generated 4.43 MHz subcarrier is gated into it, in approximately the position normally occupied by the line sync pulse. The chroma signal now passes into main mixer 1 where it beats against a locally generated carrier to produce the colour-under frequencies. Because the incoming chroma signal is at 4.433619 MHz, and the colour-under carriers are 685.546 kHz (head A) and 689.453 kHz (head B) the local signal needs to be 5.119165 kHz for head A and 5.123072 MHz for head B. As before, these local carriers are produced inside another mixer (sub-mixer 2) whose inputs consist of a 4.433619 MHz crystal oscillator, locked to off-air chrominance; and an alternating two-tone signal generated by a phase-lock loop slaved to off-air line sync.

The PLL consists of the now familiar VCO (voltage-controlled oscillator) whose output is fed to a pair of counters, one dividing by 351 (for head A) and the other dividing by 353 for head B. The appropriate divider is selected by an electronic switch, clocked at 25 Hz rate from the head drum tacho pulse. Each divider output is very close to TV line rate, and is compared with incoming sync in the phase detector, whose error output controls oscillator frequency. Thus the VCO becomes locked to 351 times line frequency during head A's scan of the tape, and to 353 times line frequency during head B's scanning period. This gives rise to output frequencies of 5.484375 MHz and 5.515625 MHz on an alternating basis, each tone lasting for one field period of 20 ms, as shown in *Figure 15.26a*. The two-tone signal is now applied to an eight-counter rendering alternate frequencies of 685.546 kHz (head A) and 689.453 kHz (head B) – see *Figure 15.26b*.

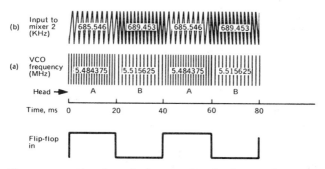

Figure 15.26. The relationship between drum flip-flop signal, VCO frequency and colour-under frequency in the Beta colour system

When this carefully arranged two-tone signal is beat against 4.433618 MHz in additive sub-mixer 2, we arrive at our 'local' signal of 5.119165 MHz alternating with 5.123072 MHz, as input to subtractive main mixer 1 – from which the colour-under signal emerges, ready for recording on tape.

Let's summarise the Betamax colour-under signal in the form in which it passes to the recording amplifier. The encoded colour signal is present, carrying colour hue and amplitude characteristics by virtue of its phase and amplitude. The swinging burst signal is also present, but both are now based on a low carrier frequency – first 685.546 kHz for the 20 ms (one field) duration of head A's recording sweep, then 689.453 kHz for the 20 ms duration of head B's operation. In addition to these, a pilot burst of constant phase is recorded by each head during the line-blanking period. The burst phases on tape are as shown in *Figure 15.23*.

Betamax chrominance replay

As may be expected, the replay process for Betamax is the inverse of that used in record. *Figure 15.27* shows the system.

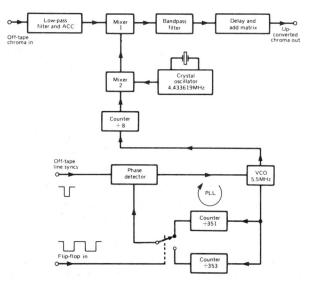

Figure 15.27. Betamax replay system block diagram showing up-conversion, line-rate de-jittering and phase restoration

The lower VCO uses the same tacho-switched divider system to produce alternate frequencies of 5.484 and 5.515 MHz for heads A and B; this time they are slaved to off-tape line sync and thus form a jittering reference. The ÷8 block presents alternating tones of 685 kHz and 689 kHz to sub-mixer 2 where they beat with a local 4.43 MHz signal to produce alternately 5.119 and 5.123 MHz. These are the frequencies required to produce a 'difference' of 4.433619 MHz when beat with the off-tape colour-under carriers of 685 kHz and 689 kHz in main mixer 1.

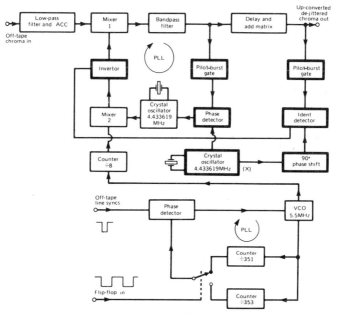

Figure 15.28. Expansion of *Figure 15.27* to show the subcarrier-rate de-jittering loop. The additional blocks are shown in bold outline

Figure 15.28 expands the previous diagram to show crosstalk compensation and subcarrier-rate de-jittering processes. The 2-line delay and matrix circuit works in exactly the same way as described for VHS and no further description is necessary. Let's examine the way in which the pilot burst is used. It will be recalled that this was inserted during record in a blank space in the chrominance signal for use as a reference. On replay, then, it is gated out of the chrominance signal after up-conversion (i.e. at 4.43 MHz) and compared in a phase detector with a reference from a local stable crystal oscillator. Any phase errors, representing jitter in the off-tape chroma signal, give rise to an error voltage input to the 4.43 MHz VCO feeding sub-mixer 2 so that its output contains a jitter component in sympathy with the off-tape signal. As before, the difference between these two jittering signals is constant, so that they cancel in main mixer 1 to provide a stable 4.43 MHz chrominance output. It can be seen, then, that like the VHS format, Betamax uses *two* de-jittering loops during chrominance replay – the off-tape line sync signal impresses timing errors on the 5.484/5.515 MHz VCO, and the replayed chroma signal itself (in the guise of the pilot-burst signal) 'slaves' the replayed subcarrier signal frequency to a stable local oscillator X by means of the 4.43 MHz PLL via sub-mixer 2.

In our study of the VHS chroma replay system we saw a similar arrangement in operation, and encountered a device (phase detector 2, *Figure 15.22*) for pulling the output signal phase within the operating range of the de-jittering PLL. The same function is performed on Betamax replay by the second pilot-burst gate and ident detector, which inverts the 5.1 MHz signal to mixer 1 when replay phase error falls outside the operating range of the loop. This inversion gives the effect of a 180° phase change on both input and output of mixer 1.

The de-jittered chroma signal, now at 4.433619 MHz, passes through a colour-killer and filter on its way to be added to the luminance signal and passed out of the machine. The pilot-burst signal, having realised its purpose of jitter compensation, is removed in the luminance-adding process.

Video-8 format chrominance

The basic principle of the colour system of Video-8 format is the same as for VHS and Betamax already described, in that the chroma phase recorded on tape is manipulated to ensure that crosstalk signals come off the tape in antiphase over a two-line period. The colour-under frequency for V8 is $(47 - \frac{1}{8}) f_h$, which is 732 kHz. It is locally generated, but in order to implement de-jittering during playback, is locked to incoming line sync in a PLL incorporating a $\div 375$ stage. We start, then, with $f_h \times 375$ = 5.86 MHz. This is divided by eight in a counter to render 732 kHz; the counter is under the influence of drum flip-flop and line-rate input pulses so that the phase of the colour-under carrier is advanced by 90° per line for head A's sweeps only, see row 1 of *Figure 15.29*. Head B's chroma record signal is not phase-changed, as row 2 of *Figure 15.29* shows.

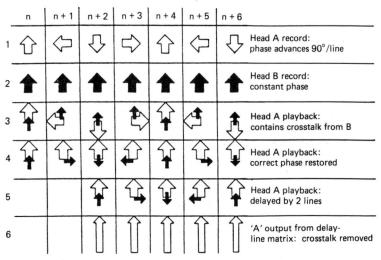

Figure 15.29. The stages in removing crosstalk interference from the Video-8 chroma signal. The 'n' numbers refer to television scan lines

During replay the chroma signals are read-out from tape according to row 3 in Figure 15.29: this slows the output from the Ch1 (A) head, and represents the main signal as recorded, together with a crosstalk contribution from adjacent B tracks, shown as small black arrows. To cancel the effect of the phase advance given to head A's chroma signal during record a corresponding phase-retard of 90° per line must be imparted to it during playback. This has the effect of restoring phase normality to the chroma *signal*, but introducing a 'twist' to the crosstalk component as shown in row 4 of the diagram. The process has not removed the crosstalk, but has paved the way for its cancellation.

Row 5 of *Figure 15.29* shows the new timing of row 4 after its passage through a 2-line delay system. Adding the delayed and non-delayed signals, so long as their timing is exactly right, gives addition of the wanted (in-phase) chroma signals, but complete cancellation of the crosstalk signals which are now in opposite phase for every line. The resultant 'clean' chroma signal is shown in the bottom row. Although we have only illustrated here the situation for a pure red signal recorded by head A, the same crosstalk-cancellation mechanism works for any colour, and also for head B's signals from which crosstalk from A tracks is removed.

The only time the system breaks down is when one line is markedly different in hue from its next-but-one neighbour. Under these circumstances of colour changes at horizontal edges in the picture the crosstalk-compensation system will introduce hue errors centred on the point of transition, the effect being worse than complete loss of crosstalk cancellation. To prevent this the chroma 2-line delay circuit is governed by a *correlation detector* which looks for large disparities in hue over a two-line period. When they are detected the effect of the chroma delay line is cancelled.

Advanced techniques are applied elsewhere in the Video-8 colour circuitry. The burst signal, a crucial reference for both VCR and TV, must be kept in the best possible condition. To give it better immunity from tape-noise its amplitude is doubled during record, and restored to normal during playback. In addition to this *burst-emphasis*, the chroma signal itself is treated for noise-reduction in a frequency-conscious amplifier which gives increasing emphasis to the outer 'skirts' of the chrominance sidebands. Corresponding de-emphasis during playback gives a useful overall reduction in 'confetti' on the monitor screen. Indeed, these record techniques ensure that most chroma signals on tape, regardless of

310

their amplitude in the 'real' signal are recorded at high levels, clear of the 'noise floor'.

Neither Video-8 nor S-VHS formats have variants to cater for Secam colour systems. Where necessary the Secam signal is transcoded to a PAL configuration during record, and back again during replay. International tape exchange is thus facilitated.

Chrominance definition of replayed pictures

We saw at the beginning of this chapter that the bandwidth of the chrominance signal is restricted during its passage through the recording system of a domestic VTR, resulting in a reduction of horizontal resolution (the ability to define vertical coloured edges) by about 50 per cent. The effect of the delay line matrices in the chroma crosstalk compensation system is to impair the vertical chrominance resolution, and the ability of the system to define horizontal coloured edges (vertical definition) is in fact reduced by a factor of eight overall. This has the effect of rendering roughly equal the vertical and horizontal chrominance definition of the display-ed picture on replay, and is considered acceptable. There is little point in maintaining good vertical resolution when horizontal definition is restricted, though this is done in the broadcast signal, where vertical resolution is governed by the number of TV lines in the picture, whereas horizontal resolu-tion depends on chroma channel bandwidth – 1.2 MHz or thereabouts in the UK system.

PAL encoding and crosstalk compensation

Although we have used the PAL chroma and burst vectors to illustrate the principle of the chrominance crosstalk com-pensation system, the PAL characteristic of the encoded chroma signal takes no part in the crosstalk compensation process. Thus the cancellation systems described above work equally well on any form of colour-encoding system. Because

311

the PAL signal (and other line-alternating colour-encoding systems such as SECAM) has a two-line pattern, the crosstalk compensation is carried out on a two-line basis to retain the swinging burst (and chroma V) characteristics. With a simple encoding system such as NTSC, a one-line crosstalk cancellation system would be possible, conferring the advantage of better vertical chrominance resolution.

16

Signal processing: audio

For several years the audio system used in domestic VTR formats depended on a longitudinal track laid along the extreme top edge of the ribbon. *Figure 13.5* shows its position, and *Figure 13.9* the placement of the stationary head which writes and reads the audio track. One of the virtues of home VTRs is their low tape consumption, but this depends on a slow tape speed, much slower for instance than an audio compact cassette. The result is a constriction on audio frequency response, particularly in LP modes where the rate of progress of the tape is halved.

The noise performance, too, of a longitudinal sound recording on video tape falls far short of other sound systems. This stems partly from the low tape speed (massive equalisation must be applied to the off-tape signal to maintain flat frequency response) but mainly from the narrowness of the tracks necessarily used. The standard longitudinal track width is 1 mm, reduced to 2 of 0.35 mm in the ephemeral stereo versions of VTRs using this system. *Figure 16.9* highlights the shortcomings of VTR longitudinal track performance, and compares them with the results obtained from the new systems to be described now.

FM audio

The ingredients for success in recording TV pictures on tape are high writing speed and FM modulation as we saw in Chapters 11 and 12. Within the immovable constraints of existing formats it was obvious that for better sound performance these virtues must also be applied to the audio signal. The obvious solution was to record an FM-modulated sound carrier in or around the helical video tape tracks; but with the vision heads and tracks already chock-full of information (see *Figures 13.29* and *15.9*) the problem was where to squeeze in the audio recording with regard to frequency-spectrum space and the 'magnetic' capacity of the tape track. Some form of multiplex system is required. We have already met this in the description of MAC TV systems in Chapter 9, where a method of getting several streams of information through a single channel path was found in time-compression and time-division-multiplex. A similar technique is in fact used in the PCM sound systems to be discussed later in this chapter. All TDM setups require signal storage at each end of the link, however. Two alternative systems emerged for home VTR formats – *depth multiplex*, used in VHS and Betamax-format models; and *frequency multiplex* in Video-8 format. The former is the most common, and will be covered first.

Depth-multiplex audio

This technique depends to some degree on the magnetic layer of the tape itself to discriminate between the video and audio signals. A separate pair of heads on the spinning drum are provided solely for the audio signal FM carriers. A typical layout of heads on the drum is shown in *Figure 16.1*. The two audio heads are mounted 180° apart and are arranged to 'lead' the video heads. The gaps cut in the audio heads have large azimuth angles: ±30°, sufficient to prevent crosstalk from adjacent tracks at the carrier frequencies (around 1–2 MHz) involved. The mounting height of the audio heads is set to place the hi-fi audio track in the centre-line of the corresponding video track (VHS) or centrally-straddling adjacent video tracks (Betamax). These track layouts are shown in *Figure 16.2*, which also gives an idea of the relative audio track widths for the two formats: in VHS the audio tracks are half the width of the vision tracks they accompany, while for Beta the video and audio helical tracks are equal in width.

Figure 16.3 gives an idea of the depth-multiplex principle. Here the heads are moving towards the right across the tape, led by the

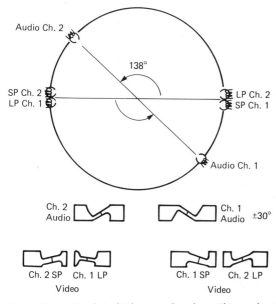

Figure 16.1. Depth-multiplex recording drum. This one has the audio heads mounted 138° ahead of the double-gapped video heads

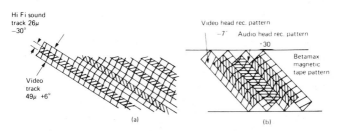

Figure 16.2. Video/sound track relationships: (a) for VHS; (b) for Betamax

audio head which has a relatively wide gap. The effect of this is to write into the tape a deep magnetic pattern penetrating some microns into its magnetic coating. Shortly following the audio head comes the regular video head, whose gap is in the region of

0.25 micron. Writing (for the most part) at higher frequencies, it creates shallower, shorter magnetic patterns which penetrate less than 1 micron into the magnetic surface. Thus the recorded tape (top LHS of *Figure 16.3*) contains a 'two-storey' signal: a buried layer of long-wavelength audio patterns under a shallow top layer of video patterns.

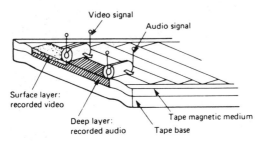

Figure 16.3. Depth-multiplex principle: separate magnetic layers are written into the tape

During replay the same heads operate on the same tracks as before. The video head picks off its track with little impairment, and with minimum crosstalk from sound tracks because of the large disparity (36° for VHS-SP) between the azimuth angle of recorded track and replay head. The audio head during replay is handicapped by the barrier presented by the 'video layer' on tape, but the resulting 12 dB or so of attenuation - thanks to the use of bandpass filters and the noise-immunity of the FM carrier system – does not prevent noise-free reproduction of the baseband audio signal as long as the tracking is reasonably correct. For hi-fi sound the tracking performance is critical, especially in LP modes where narrow tracks tend to magnify any errors.

As with the video heads, switchover between the two rotating audio heads is carried out during the overlap period when both heads are momentarily in contact with the tape, one just leaving the wrap and one just entering. Because of the angular offset between video and audio heads on the drum (*Figure 16.1*) the head switchover point for the latter is set by a second, *delayed* head flip-flop squarewave triggered from the drum tacho-pulse. The FM carrier signals to and from the audio heads on the drum require their own rotary transformers, which may be concentric with those for the video signals under the drum; or mounted above as in the photo of *Figure 16.4.*

Figure 16.4. Head drum mounting in a Panasonic Hi-Fi VHS machine. The rotary transformer for audio signals is above the drum

The frequencies used for stereo hi-fi in depth-multiplex systems are 1.4 MHz (L) and 1.8 MHz (R) for VHS; 1.44 MHz (L) and 2.10 MHz (R) for Betamax. In all cases the FM modulation sidebands extend for about 250 kHz on each side of the (unsuppressed) carriers. Each audio head deals with *both* FM carriers throughout its 20 ms sweep of the tape – during replay the carriers are separately intercepted by bandpass filters for processing in their own playback channels.

Frequency-multiplex audio

An alternative approach to helical sound recording is to use the *video heads* to lay down on the tape an FM audio soundtrack. It is only possible where a single (monaural) audio track is required, with a baseband frequency response limited to about 15 kHz. As *Figure 13.29* shows the audio FM carrier is based on 1.5 MHz and has a deviation-plus-sideband width of about 300 kHz. It is just possible to squeeze this 'packet' between the outer skirts of the upper chrominance and lower luminance sidebands, permitting it to effectively become part of the video signal so far as the heads and tape are concerned. The signal FM audio carrier is added to the FM luminance signal in the recording amplifier, and laid on tape at a low level – some 18 dB below that of the luminance carrier. This suppression of sound carrier level helps prevent mutual interference between sound and vision channels, whose outer sidebands overlap to some degree.

In replay mode the separation of the various off-tape signals in *Figure 13.29* is carried out by four separate filters: a low-pass type with cutoff around 180 kHz for ATF tones; a low-bandpass filter centred on 732 kHz for colour-under signals; a narrow bandpass one tuned to 1.5 MHz for interception of the audio-FM carrier; and finally a high-pass (roll-on about 1.7 MHz) acceptor for the FM vision signal with its sidebands.

Audio electronics

In 'longitudinal' audio systems the only processing necessary for the audio signal itself is some recording equalisation and the addition of a bias signal as described in Chapter 12. For hi-fi recording a great deal more work must be done on the input signal to condition it for multiplex recording, and to reduce the overall system noise level to a point far lower than can be achieved with conventional sound recording systems on tape or disc. The techniques used are almost identical for depth-multiplex and frequency-multiplex systems; the main differences lie in the practicalities of the formats used, i.e. VHS hi-fi requires duplicate processors to handle left- and right-channel audio signals routed to and from the separate rotary audio heads, while the 'AFM' section of a Video-8 format VTR has but a single channel to deal with via the existing video heads. In all cases much of the audio processing circuitry is common to both record and replay.

Figure 16.5 shows, in generalised form, the arrangements for a

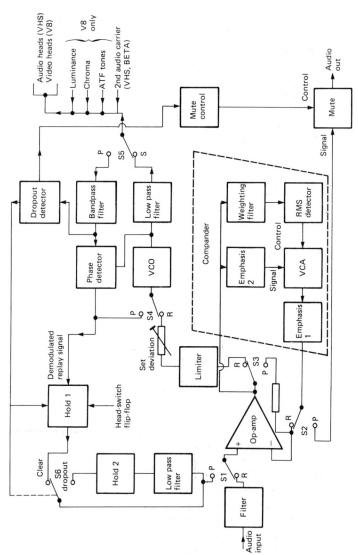

Figure 16.5. Block diagram of hi-fi audio signal processing, fully explained in the text

319

single channel of hi-fi sound processing. In record the audio signal (left of diagram) is first passed through a high-cut filter to remove supersonic frequencies, then via switch S1 to the non-inverting input of an operational amplifier. The op-amp output (ignoring for a moment S3) takes two paths: one through a pre-emphasis 2 network to the input of a VCA (voltage controlled attenuator); and one via a weighting filter to precision RMS-level detector. The output of this device reflects, from moment to moment, the true RMS (root-mean-square) amplitude value of the signal it sees. The RMS detector output is now applied as a control potential to the VCA. Thus the output from the VCA is large for high-level audio signals, small for low level signals. It is routed via S2 to the inverting input of the op-amp. The effect of this variable negative feedback is to 'compress' the dynamic range of the audio signal emerging at the op-amp output. The compression ratio is 2:1, expressed in decibels on the left-hand side of the diagram in *Figure 16.6*. The amplitude-compressed signal, since it will be conveyed by an FM carrier, has carefully-tailored pre-emphasis characteristic imparted by the emphasis-1, emphasis-2 and weighting filter sections in *Figure 16.5*.

Continuing with that diagram, the compressed and pre-emphasised audio signal now passes through S3 to a limiter whose purpose is to clip any signal excursions that may cause over-deviation in the FM modulator. Next follows an attenuator for pre-setting, via S4, of FM deviation in the VCO which performs frequency modulation. The FM carrier is conveyed by S5 to a low-pass filter whose output is added to ATF, chrominance and luminance-FM recording signals for passage to the video heads (V8 format); or passed directly to the rotary audio heads in VHS-format machines.

As is obvious from *Figure 16.5* all the important processing blocks are used again during playback. Replay head input is now routed by S5 to a bandpass filter, centred on 1.5 MHz for Video-8, 1.4 MHz for VHS L-channel, 1.8 MHz for VHS R-channel etc. With its loop completed by S4 the VCO now forms part of a PLL detector whose demodulated output signal goes via S6 and S1 to the non-inverting input of the main op-amp, the gain of which is now set by the fixed feedback resistor from S3 to the *inverting* input.

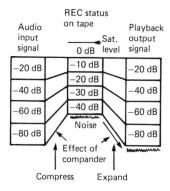

Figure 16.6. Overall effect of companding on the audio signal. The dymamic range is greatly compressed for the 'on-tape' phase

Again the op-amp output signal is passed via the emphasis 2 filter to the VCA; again the latter is under the control, gain-wise, of the RMS detector whose input is conditioned in turn by the weighting filter. Once more the output of the VCA passes through the emphasis-1 circuit, now to form the audio output signal via S2 This time, however, the entire circuit within the dotted outline in *Figure 16.5* is in the direct signal path rather than part of a negative feedback loop as it was during record. As a result its operation is reversed to 'mirror' its previous functions. Thus purely by the action of S2 what was a compressor/pre-emphasiser has become an expander/de-emphasiser, using all the same filters and components. Hence the expression *compander* and the symmetrical appearance of the diagram in *Figure 16.6* whose right-hand side shows the effect of expansion of the signal's dynamic range during replay.

The most significant benefit of the companding circuit may not be obvious as yet. Plainly, it is advantageous to boost the low-level components of the audio signal so that they *significantly* deviate the FM carrier to swamp noise. The 'tape-noise' in an FM system is typically −45 dB, but since the lowest-level signals coming off tape are effectively 'attenuated' during playback, this −45 dB noise floor is depressed to −80dB in practical systems. This ingenious combination of FM and companding techniques permits us to achieve an overall dynamic range and S/N ratio better than that inherent in the tape system itself as defined in Chapter 12 and *Figure 12.7*.

Dropouts can seriously upset the operation of a hi-fi audio system, and masking of these is even more important than for video. In *Figure 16.5* a dropout detector (top right) monitors the off-tape audio FM carrier and throws switch S6 whenever it sees one, whereupon the 'hold 2' circuit and associated low-pass filter provide dropout compensation. Hold 1 circuit is primarily concerned with masking the dropout and disturbance associated with head-switching. If the dropout count becomes too high (i.e. severe mistracking) a mute circuit comes into operation – on VHS and Beta the audio output now reverts to the longitudinal sound track(s) which are always provided as back-up, and to confer compatibility with other machines and tapes.

PCM audio

A brief outline of the method of quantising analogue signals was given in Chapter 9, starting on page 168. All newly-developed hi-fi audio systems use digital encoding; examples are compact disc, the Nicam TV stereo plan, DAT tape recording and MAC TV sound transmissions. Although digital transmission and recording requires more bandwidth than other systems it has the advantages that the two-state signal is more robust than its analogue counterpart, and (so long as the quantising rate is high) is capable of superb signal/noise ratio and wider dynamic range.

The Video-8 format has provision for stereo PCM in addition to the monaural AFM sound facility described above. The most expensive V-8 camcorders and homebase VTRs are fitted up for PCM operation. It is impractical to give full details of the system in this book, and what follows is a basic outline of the technique used.

Figure 16.7 shows the functional blocks in the PCM sound processing. Initially the audio signal for record is amplitude-compressed in a compander similar to that in *Figure 16.5*. Next follows a quantisation process in which it is sampled at $2 f_h$, (31.25 kHz) to 10-bit resolution. Since the tape system here

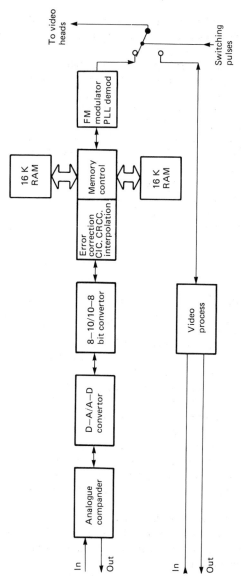

Figure 16.7. Outline of PCM audio recording system

323

cannot cope with 10-bit data a conversion is carried out to 8-bit: the process is a non-linear one, in which low-level signals are in effect given 10-bit (1024-level) descriptions, falling in three stages – as signal level increases - to 7-bit data (128-level) for the largest signal excursions. This has the effect (during playback) of concealing quantising noise in the loudest sound peaks, where they go virtually unnoticed; indeed the overall signal-to-noise ratio is subjectively equivalent to 90 dB.

Unless their transmission/recording media are very secure, digital data systems need error-correcting artifices to repair or conceal corruption of the data by noise and distortion. About 38.5 per cent *redundancy* is imparted to the PCM audio data by the addition to it of further data in the form of a cyclic redundancy check code (CRCC) – used during playback for corruption-test and 'first-aid' purposes. The data rate is high, and the effect of a tape dropout would ordinarily blow a hole in the information stream; to prevent this the data is 'scattered' on tape according to a cross-interleave code (CIC), part of the Video-8 format. The effect of a tape dropout thus becomes distributed during replay, and the 'frayed edges' can be repaired by use of the CRCC and an additional parity check system.

The 8-bit data words are now temporarily stored in a pair of 16 K RAM memories. Writing to memory is performed in real time. Readout from memory is much faster: all the data (which contains information on both stereo channels) is clocked out in less than 3 ms at 20 ms intervals. The effect of this 7:1 time-compression is to push up the data-rate to about 2 Mbits/second, but to confine the audio transmission period to a small time-slot, in very similar fashion to that of the MAC system in *Figure 9.5*. As we shall see in a moment, a separate place is found on the videotape for this data-burst. The two-state data signal is tone-modulated at 2.9 MHz for data 0 and 5.8 MHz for data 1; this is called frequency shift keying (FSK).

At this point the PCM signal is ready to go onto tape. The switches on the right of *Figure 16.7* change over once per field period, in synchronism with TV field rate and RAM readout. This feeds data to the video heads alternately, during a period when each is scanning a 'forward extension' of the helical vision track on tape – this is illustrated on the right of *Figure 16.8*. The

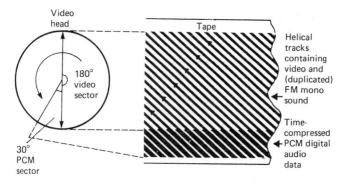

Figure 16.8. The PCM data is recorded on a 'forward-extension' of the video tape tracks; an extra 30° of head rotation is reserved for this

conventional video tracks (which also contain AFM audio information, duplicating the PCM sound track) are recorded over 180° of the head track, but the head/tape angle is such that they occupy about 5.4 mm of the 8 mm tape width. The extra 30° or so of tape wrap shown on the right of *Figure 16.8* is devoted to PCM recording: while one head is writing PCM data, the other (diametrically opposite on the drum) is recording the last lines of the TV picture at the top of the *video* tracks on the right of the diagram.

During playback, head switching ensures that PCM data read off the tape is routed to the audio section for the appropriate 30° scan/3 ms. As in *Figure 16.5* the record VCO FM modulator is now switched to perform PLL demodulation, producing binary data at its output. Continuing to the left in *Figure 16.7*, the data, still in time-compressed form, is read into the same pair of 16 K memories as were used during record.

Memory readout takes place in 'real time', expanding the data to give *continuous* data at a lower bit-rate – that at which the memories were loaded during record. Readout sequence is governed by the CIC (see above) in order to de-interleave the data, scattering and fragmenting errors in the process. In the memory control chip is also carried out error correction by means of the CRCC and parity checks mentioned earlier. For dropouts too severe to be repaired by these means, *interpolation* provides a 'patching' system which, in the face of sustained and continued

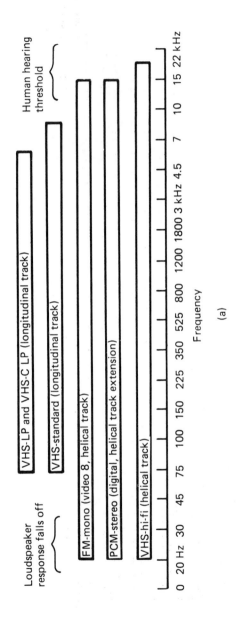

(a)

326

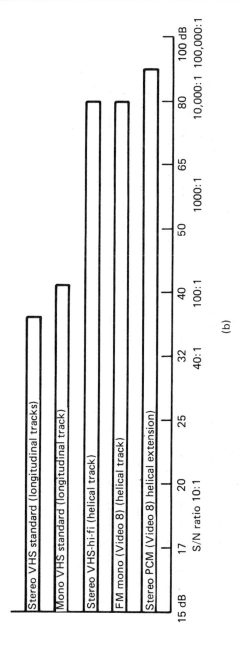

Figure 16.9. Audio system performance of various formats. Bar-chart (a) illustrates frequency response capability; bar-chart (b) shows signal to noise ratio. The upper two bars in (b) are shown in 'raw' state, i.e. without the sophisticated noise-reduction systems usually applied

corruption, devolves to a PCM mute action, switching the audio output line back to the AFM sound track. If this is also corrupted, silence will ensue!

The 8-bit 'reconditioned' data now passes to the 10-bit conversion stage to make ready for D-A conversion. As is common in these designs the record A-D convertor is used for this, now switched to perform D-A operation. The analogue signal reconstituted at the D-A convertor output is still in amplitude-compressed form, and is now expanded to full dynamic range in a logarithmic compander – the dotted-box section of *Figure 16.5*.

The relative performance of the audio systems described in this chapter is charted in *Figure 16.9*, where (a) shows the frequency responses, and (b) the S/N ratio capabilities.

17

Servo systems and motor drive

In an audio tape recorder, we have a single drive system to rotate the capstan, and this pulls the recording tape past the sound head at a constant, fixed speed. Provided this speed is the same during record and playback, the programme will be correctly reproduced, and if the tape speed is arranged to conform to the standard (4.75 cm/second for audio cassette machines) we have the further advantage of interchangeability between tapes recorded and played back on different machines. The only requirements, then, of a drive system in an audio tape recorder are that it should run at a specific speed, and that short-term speed variations (wow and flutter) are kept at a low-enough level to prevent noticeable changes of pitch in the reproduced sound. This simple drive system is possible because an audio waveform carries *all* the necessary information about the sound signal at any one instant, and this waveform is laid down along the tape as a single longitudinal stream of magnetic patterns.

The situation with TV signals is rather different. Because the TV image is a two-dimensional display, and only a single transmission channel is available, the image has to be built up by line-by-line scanning of the scene, with 312½ lines making up one field, and taking 20 ms to do so, and two fields interlacing to form a complete frame, or picture. Thus for television signals it takes 40 ms to build up one picture, and we have introduced a second factor into the signal: time. The video signal, then, has built-in markers to indicate this second dimension of time, and the markers which signify the beginning of each new field every 20 ms are very important in video tape recording, as we shall see.

Another factor in the process of recording TV signals, which only has an audio-recording counterpart in DAT machines, is the idea of laying down parallel slanting tracks on the tape by means of a rotating head drum. If replay of these tracks is carried out in a random manner (even at the correct speed) there is no reason why the replay heads should follow the path taken by the heads during record, so that each replay head may well read between the lines, as it were, and reproduce a signal consisting of a mixture of the information from two adjacent tracks. We have already seen that adjacent tracks contain different video signals recorded at different azimuth angles, so that this form of mistracking will make a noisy mess of the reproduced signals.

To overcome these problems, a very precise control system, known as a servo, is required to govern the head drum and capstan drive systems. In this chapter we shall examine the purpose and mode of operation for both head drum and capstan servos, and see how they govern the relationship between video head position and tape track placement.

Requirements during record

We have seen in Chapter 13 that two video heads are used in home VTRs, each of which lays down one complete field of 312½ TV lines during its scan of the tape. At some point during the field period one video head has to 'hand over' to the other, and as this will cause a disturbance on the reproduced picture, the 'head interchange' or switching point is arranged to lie at the extreme bottom of the picture, just before the field sync pulse (*Figure 13.16*). This means that the disturbance will be just off the screen in a correctly set-up monitor or TV set due to the normal slight vertical overscan. This means that during record we need to arrange for each head to start its scan of the tape about half a millisecond (corresponding to approximately eight TV lines) before the field sync pulse occurs in the signal being recorded; and hold this timing, or phase relationship, steady throughout the programme.

If this relationship were not present, the head-switching disturbance effect would move into the active picture area, and drift up or down the screen. The mechanics of the deck arrangement and the position of the 'entry guide' determine the point at which the moving tape starts its wrap of the head drum, as shown in Chapter 13. The electronic servo loop sees to it that each head passes this point and starts to write its field of information just before the field sync pulse occurs in the video waveform being recorded. To achieve this, we need some marker to tell the servo system the angular position of the head drum. There are many possible ways of generating this, but the most common is the use of a small permanent magnet fixed to the underside of the head drum, and arranged to induce a pulse in a nearby stationary pickup coil at each pass of the magnet, as shown in *Figure 17.1.*

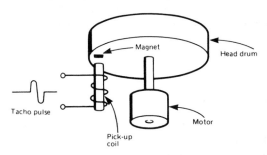

Figure 17.1. Basic tacho arrangement with rotating magnet and pick-up coil

Other methods of generating these pulses are possible. In some VTRs a drum-mounted 'flag' interrupts the light-path of an *opto-coupler* once per revolution; or *Hall-Effect* devices are used. However it is made, each time this *tacho pulse* appears we know the head drum is at a certain angular position.

The tacho pulse, then, may be regarded as a phase reference for the head drum. If the tacho pulse is applied to a phase detector for comparison with the timing of the incoming field sync pulse, an output voltage will be produced (see page 278 proportional to the timing error between these two input signals. If this error voltage is applied to the head drum drive motor, the control loop will ensure a fixed relationship between the field sync pulse and the angular position of the head drum – yet another example of a PLL, but this time with a mechanical system within the loop. This simple servo system is illustrated in *Figure 17.2.* By delaying either the field

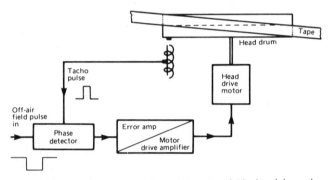

Figure 17.2. Head drum servo loop during record. The head drum phase (angular timing) is locked to off-air field syncs

sync pulse or the tacho pulse on its way to the phase detector, we can set up any desired phase relationship between incoming field sync and video head position, governed by the length of the delay introduced. It is arranged that the phase relationship is such that the video tracks are laid down with all the field syncs at the bottom of the tape, as in *Figure 13.15.* The fact that one 20 ms TV field is recorded by each head during one half-turn of the head drum precisely dictates the rotational speed of the drum – it comes out at 40 ms per revolution, or 1500 rev/min.

What is required of the capstan servo during record? As we have seen, it takes no part in determining the timing of each one-field track, which is wholly the province of the head drum. The capstan speed governs the spacing of the video tracks, however, and as we know, they are laid down so that they just abut each other; this calls for a steady and even flow of tape past the head drum. The audio signal is recorded longitudinally on the tape in the normal way, so good sound and jitter-free pictures depend on a steady capstan speed. The actual capstan speed depends on the format, being 2.34 cm/second in VHS, for example. During record, then, the capstan needs only to be held at a constant speed, and in some early VTR designs no capstan servo was used. All current VTRs employ a capstan servo, however, which on record is slaved to a high-stability frequency reference known as a *clock*. This may be a stable crystal oscillator, 50 Hz mains frequency or the field sync from the signal being recorded.

The control track

When the video tape being recorded is played back, we will need a reference signal to define the physical position of each recorded video track on the tape. The situation is akin to that of cine film, where sprocket holes are provided to

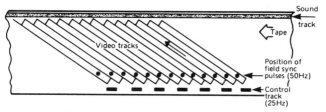

Figure 17.3. Configuration of tape track pattern as a result of the action of the servo loop in *Figure 17.2*

ensure that each frame is exactly in position in the film gate before the projector's shutter is opened. Our video tape 'sprocket-holes' take the form of a control track recorded along the edge of the video tape by a separate stationary head just like the sound head. The control track consists of a 25 Hz squarewave pulse train. One pulse, then, is recorded for each two fields and the track pattern is shown in *Figure 17.3* which also shows the position of the sound track. We shall see later how the control track is used during playback.

Servo operation during playback

Let's assume we load into the VTR a good pre-recorded tape, and initiate the replay mode. To get a stable and noise-free picture we need the same close control over the capstan and head drum drive that takes place during recording, but the control references are somewhat different. We'll take the operation of the head drum servo first. The head-changeover point, which needs to correspond with that used during recording, is controlled during replay by the head drum tacho pulse, which triggers the head-switching bistable via a variable time delay. The delay is adjusted so that head switching takes place at the correct instant, just into each tape track.

The speed of the head drum during replay determines the line and field frequencies of the off-tape video signal. Thus the drum needs to spin at precisely 1500 rpm as during

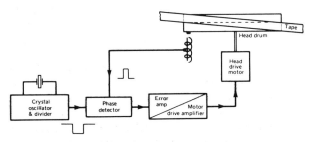

Figure 17.4. Head servo on replay, locked to stable local reference

record, and this is achieved by slaving it to a stable frequency reference – again 50 Hz mains or a crystal reference is used. Phase-locking takes place in the same phase detector as was used during record; the positional feedback input to this is again the tacho pulse from the drum pickup coil, but the reference input to the phase detector now becomes the local high-stability frequency source – a crystal oscillator or the 50 Hz mains supply, as shown in *Figure 17.4*.

During replay we have to ensure that each video head scans down the centre of its intended track, and this is achieved by adjusting the lateral position of the tape around the head drum to line up the head scans with the recorded tape tracks, and hold them constantly in alignment with each other. This calls for another servo operation, and a moment's thought will show that this 'lateral lining-up' procedure can be carried out by adjusting tape position (via the capstan servo) or head drum position (via the drum servo). In either case we need a reference signal related to the position of the video tracks on the tape, and this is where the control track, laid down during record, comes into use. We'll describe the capstan-tracking system, as used in most of the early machines.

In addition to its duty of pulling the tape through the machine at a steady speed equal to that used on record, the capstan is now required to position the tape relative to the head drum to achieve correct tracking. Although the video tape tracks are very narrow, they are at such a shallow angle (say 5°) to the tape path that a relatively large lateral movement is required to 'move' the tape across from one track to the next. Let's see how this drum/tape-position phasing is effected.

During playback the control track head reads the 50 Hz pulse train off the tape to provide a 'track position' reference signal for the capstan servo. This is compared in a phase detector with the same stable local signal (50 Hz mains or the frequency-divided output of a crystal oscillator) to which the head drum is slaved, and any error is fed to the capstan motor to pull it into lock. Thus a fixed relationship is set up between drum position and tape track placement, and the phasing, or timing, of this relationship can be adjusted by including a variable delay in the path of the control track signal to the phase detector. This variable delay is set by the tracking control, and it is adjusted on playback for minimum

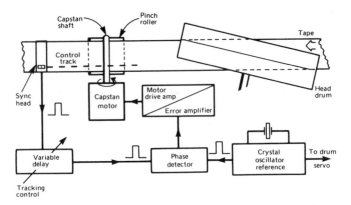

Figure 17.5. The capstan servo during replay, locked to control track pulses

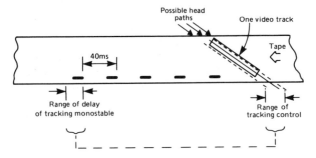

Figure 17.6. The effect of the tracking control on replay. It 'phases' the lateral tape position to line up video tracks with the path of the vision head

noise in the picture. signifying optimum tracking. A diagram of this system appears in *Figure 17.5*, while the effect of adjusting the tracking control is conveyed in *Figure 17.6*.

Some VTRs carry out the replay tracking function via the head drum servo; those machines which have no capstan servo have to! In all cases, the requirements and principles are the same.

Automatic tracking systems

So long as the physical position of the sync head is correct, with reference to the entry-point of the tape onto the drum in the recording machine, its tapes will replay in any similar machine, provided that the latter is also in good mechanical alignment. The tracking control only operates during replay, and is only required to take up tolerances due to mechanical wear, tape tension variations etc. It is a user control, and maladjustment often impairs picture performance.

Various forms of automatic tracking system have been developed to overcome this problem: we have examined the ATF arrangements of the Video-8 format, which are based on the *DTF* system used in the now-obsolete V2000 format. With these no manual tracking control is necessary or provided since 'tolerance' errors, both long term and short term, are automatically taken up. Any errors due to tape-path anomalies between record and playback machines are similarly removed. In VTRs using longitudinal control tracks a form of automatic tracking control is possible. The off-tape FM video carrier's *amplitude* is independent of signal content, depending only on the efficiency of tape-head transfer. This means that tracking accuracy can be measured by monitoring the replay FM 'envelope' amplitude. Auto-tracking systems operate by varying the phase delay (bottom left-hand corner of *Figure 17.5*) to maximise replay head output. This can be done on a continuous or a 'stab and store' basis.

Servo basics

We have seen what servo systems do in a VTR and why they are necessary. Time now to find out how they work! All servos are closed-loop devices, where the 'end result' is fed back to modify the 'processing'. Perhaps the simplest example of a closed-loop system is an electric immersion heater, where the water temperature is monitored by a thermostat which switches the heating element on and off to maintain a reasonably constant temperature regardless (within the capabilities of the heating element) of the demand being made on

337

the hot water supply. A more scientific example of a closed-loop system is found in a TV receiver where incoming line-synchronising pulses are used to hold in step the horizontal scanning lines of the picture. A sample of output frequency and phase is fed back to the flywheel sync circuit and its timing compared with that of the incoming sync pulses. Any discrepancy gives rise to an error signal, and this, after filtering, controls the frequency of a VCO. The local frequency and phase are pulled into lock with the controlling, or reference pulses, and the oscillator is then said to be slaved to the reference. In discussing colour-under systems and colour crosstalk compensation processes we met the closed loop (PLL) system several times, and the phase discriminator and VCO at the heart of it are discussed on pages 278-9.

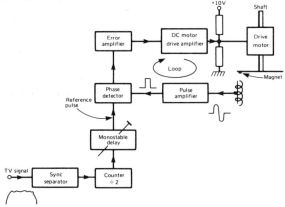

Figure 17.7. Basic servo loop. The reference signal is incoming field sync pulses

Figure 17.7 shows the basic servo loop. In this case the motor is a permanent-magnet DC type, so that its speed is proportional (for a given mechanical load) to armature current. On its lower shaft a disc with a magnet attached so that one pulse is induced in the tachogenerator coil for each revolution of the shaft. This pulse, at 25 Hz, is timed and

338

shaped in the pulse amplifier and then fed to the phase detector. A second input to the phase detector forms the reference, and in our example it is derived from the field sync pulse of a video signal via a sync separator, divide-by-two counter and monostable delay, whose function will be explained shortly. Assume the DC-coupled motor drive amplifier is arranged to have a standing 5 volts DC at its output with zero input voltage, and that the motor is designed to rotate at about 1500 rpm with such an input voltage. This rotational speed will give rise to tacho pulses at 25 Hz, and if these are coincident in time with the reference pulses, the phase detector will provide zero output and the motor will continue to rotate at 1500 rpm. Now, if the mechanical load on the motor is increased, its speed will tend to fall, and the phase detector will start to see a time-lapse between the arrival of a reference pulse and that of the tacho pulse, giving rose to an error voltage into the DC amplifier. This error voltage then drives more current through the motor and so increases its speed to restore the status quo. When reference and tacho pulses again become coincident in time, steady-state conditions are restored in the servo loop. Should a speeding-up of the motor occur, the circuit operates in the opposite way to correct this.

Sample and hold circuit

Because of the relatively low pulse repetition rate involved in VTR servos, and the degree of precision required of them, a special type of coincidence detector is required, known as the sample-and-hold phase detector. A block diagram to illustrate the principle is shown in *Figure 17.8*. The feedback pulse from the tacho-generator is made to initiate a linear voltage ramp, or sawtooth, symmetrical about zero potential as shown in the waveform. This ramp is applied to a gate (in the form of an electronic switch) which is normally held closed. When the reference pulse comes along it momentarily opens the gate and permits a *sample* of the ramp to charge the sampling capacitor C. Let's assume the reference pulse

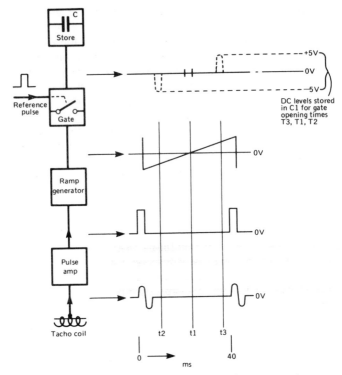

Figure 17.8. Sample-and-hold phase detector with timing waveforms

appears at the instant when the ramp is passing through zero volts, i.e. at time t1. When the gate is briefly opened at this time C will 'see' zero volts and acquire no charge. If the ramp and reference drift apart in phase so that the gate is opened at time t2, C will briefly see a negative potential, to which it will charge. Similarly, if the reference pulse comes late (t3), C will see a positive potential when the gate opens. Provided the discharge path or load on C is sufficiently high it will retain its charge between gate pulses, and that charge will indicate by its polarity the *direction* of the timing error, and by its magnitude the *amount* of the timing error. By amplifying and smoothing the voltage charge of C we shall arrive at an error voltage which can be used to control the speed and phasing of a drive motor.

If the circuit is going to have a symmetrical operating range, the ramp needs to be linear. The simplest way to generate a ramp waveform is to allow a capacitor to charge towards a fixed potential via a resistor as in *Figure 17.9a*. The problem with this simple circuit is that as the capacitor acquires charge, the voltage across it increases, leaving less for the resistor. The current in a resistor is proportional to applied voltage, so that the capacitor-charging current decays with time, giving rise to the non-linear ramp shown in *Figure 17.9b*. To ensure a straight ramp it is necessary to charge the capacitor at a constant current, and this can be achieved in two ways.

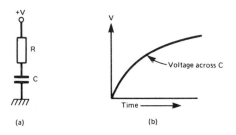

Figure 17.9. Ramp generator: (a) shows a simple CR charging circuit, and (b) the resulting exponential charging curve

The simplest is to charge the capacitor from a high voltage source via a high-value resistor. Provided the charging process is stopped (by closing a switch across the capacitor) before the capacitor has charged to a small fraction of the total voltage, the ramp appearing across it will be linear enough for practical purposes. This idea was embodied in many valve TV receivers, where the triode section of the field time base acted as a switch across the ramp generating capacitor. The latter charged a small way towards the 'boost' potential of 800 V or so via a resistor of about 2 MΩ. The second method of constant current charging is by means of a feedback circuit, of which the *Miller Integrator* and the

Bootstrap circuit are the foremost examples. Let us examine the latter type.

Bootstrap configuration

A basic bootstrap circuit appears in *Figure 17.10*. The ramp capacitor C1 charges towards the +ve voltage line via R1 and R2. In doing so it raises the base voltage of emitter-follower Tr1 and a replica of the charging ramp appears at Tr1 emitter,

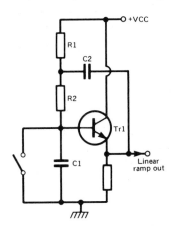

Figure 17.10. The bootstrap ramp generator. This circuit can generate a linear ramp

now at low impedance due to the current amplification of the transistor. The 'followed ramp' is transferred to the top of R2 by the large capacitor C2, so that the charging resistor R2 has a similar ramp waveform at each end. As a result the voltage across it does not change as C1 charges, and the current through it is constant. A constant charging current makes for a linear ramp, and this is taken off at low impedance at TR1 emitter. To start the ramp, the switch across C1 is opened; when the switch is closed the ramp will be terminated. In practice this switch is a transistor, turned on and off by the tacho pulse.

There is no reason why the tacho pulse should initiate the ramp and the reference pulse open the gate in a sample-and-hold detector, and the roles can be reversed. The important thing is that the sampling capacitor will store a voltage proportional to the phase error between the two. In practical circuits both configurations are used.

Monostable delay

As we have described, the phase detector in a servo system will operate to achieve phasing of a rotating shaft to a reference pulse. If we need to adjust the phase relationship between the two, we must introduce a variable time delay in the path of one of the pulse inputs to the phase detector.

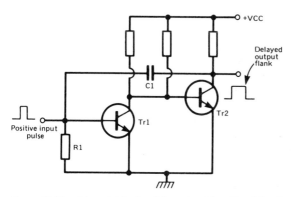

Figure 17.11. Monostable circuit for pulse delay. The delay depends on time-constant C1, R1

Because the pulse in either case is a timing reference rather than a specific waveshape, it is not necessary to use an analogue delay line; a monostable transistor pair can be used to produce an output pulse at any desired time after triggering pulse, the time delay being governed by an RC time constant, as in the circuit of *Figure 17.11.* Tr1 is normally off due to its having no base current feed. The arrival of a positive pulse at its base will turn it on, robbing Tr2 of base current. Tr2 now switches off and its high collector voltage pulls up both plates of C1, sustaining base current in Tr1. C1

343

will begin to charge via R1, and when the voltage across R1 has decayed to the point when base current can no longer be sustained in Tr1, the circuit suddenly reverts to its stable state with Tr1 off and Tr2 on. The negative spike generated at this moment on Tr2 collector forms the delayed output pulse and if necessary can be inverted to form a positive pulse. By making part of R1 variable the monostable time delay can be adjusted, and the servo phasing with it.

Motor speed control

The drive motors used for head drum and capstan may be of several different types. Early VTR models favoured a miniature precision DC motor with permanent magnetic field. Its speed is proportional to armature current and for this reason it lends itself well to servo control. Another early approach was to use a large mains-powered synchronous motor with a design speed slightly faster than that required. Here the servo output stage controls DC current in a braking coil whose magnetic field embraces a copper or aluminium disc on the motor shaft. Here eddy currents, proportional to the current in the braking coil, load the motor and reduce its speed. The system was heavy in terms of power consumption and weight, and is now obsolescent.

A third alternative is the HF synchronous motor. This AC type rotates in synchronism with the frequency of its drive voltage; and for servo control would be fed from the amplified output of a VCO under the influence of the error voltage from the servo's phase detector.

Far the most popular now are direct-drive motors. Here the rotor consists of a disc magnet into which are 'printed' a number of N and S poles. The stator is formed by a number of coils (usually six) arranged in a radial pattern, and fed by an IC multiphase driver circuit. The current-pulse timings are governed by Hall-type magnetic sensors, and speed/torque by stator coil current, in turn controlled by the voltage applied to the drive IC, often built onto the motor itself.

Motor coupling

First-generation home VTRs used belt drive to the 'operating' shafts of the capstan and head drum. Additional jockey wheels and slipping clutch arrangements were provided to drive the tape reel turntables from one of the servo-controlled motors, not only during forward mode but also during fast forward and rewind operation. The ultimate in motor economy was reached in an early Betamax type by Sony, model SL8000UB, in which head drum, capstan, tape reel drive (forward, fast forward and rewind) and even tape threading were all carried out by a single large AC motor with an eddy-current braking servo system.

Current machines tend to use direct-drive DC motors in which the capstan and head-drum spindles are actually the motor shafts. This greatly simplifies the deck layout and also reduces maintenance by eliminating wear-prone belts and friction wheels. Closer control and greater versatility are also possible with this approach, which can involve up to five purpose-designed motors to drive separately the head drum, capstan, tape reels and threading mechanism. In front-loading machines yet another motor may be employed for cassette transportation.

Practical servo circuits

Current VCRs of all formats use ICs in their servo departments, and the circuit diagrams for these are little different to the block diagrams we have studied so far. To illustrate the working of a ramp generator and sample/hold circuit we must turn to earlier designs, one of which appears in *Figure 17.12*. Here we have a reference signal, in the form of a 25 Hz rectangular wave entering on the left. During each positive half cycle D1 is conductive and effectively short-circuits the charging capacitor C1. For the 30 ms or so duration of the negative portion, however, D1 is off and C1 is allowed to charge via R1. TR1 and C2 provide bootstrap action and a linear ramp is thus generated at point A, whose slope is

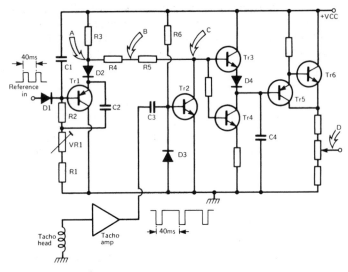

Figure 17.12. Ramp generator and sample-and-hold detector for VTR servo

adjustable with VR1. The 'ramp down' waveform appears at point C, where it is grounded in T2. This transistor acts like a switch, normally held on by base bias via R6.

The tacho generator coil at the bottom of the diagram generates one pulse for each revolution of the head drum or capstan, as the case may be. This pulse is shaped and amplified to appear as short-duration negative spikes at C3. Each time one appears, Tr2 is momentarily switched off, and its shunting effect on the ramp waveform removed; a sample of ramp voltage *at the instant of the gating pulse from the tachogenerator* is allowed to pass onto the bases of TR3 and TR4. If this sample is higher than TR3 emitter voltage, the transistor will conduct and charge the storage capacitor C4 to the sampling level. If, however, the requirement (due to direction of phase error) is such that charge needs to be 'pulled out' of C4, TR3 and D4 will remain reverse-biased and non-conductive, but TR4 will come on to discharge C4 to the new sampling level.

For C4 to maintain its charge between sampling pulses it needs to be looking into a high impedance, and this is provided by the compound emitter follower Tr5/Tr6, the output of which (point D) forms the error signal.

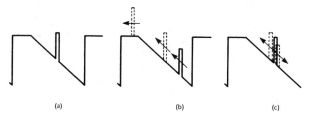

Figure 17.13. The waveform at point B of *Figure 17.12.*(a): correct operation; (b); servo unlocked, pulse running 'backwards' up ramp; (c): pulse jittering about correct point due to mechanical problems

The junction of the two 5K6 resistors R4/R5, point B, is the measuring point for the ramp and sample circuit. An oscilloscope connected here will show the sample pulse sitting on the ramp, and if the circuit is correctly adjusted the pulse will appear half-way down the ramp as in *Figure 17.13a*. This will offer a symmetrical correction range. If the servo is out of lock, due perhaps to a fault condition, the pulse will appear to run up or down the ramp at a speed proportional to the frequency/rpm error, as conveyed in *Figure 17.13b*. Mechanical jitter, due to worn bearings or faulty friction drives, causes the pulse to jitter about the correct point on the ramp, as in diagram (c).

Motor drive amplifier

The amplification system between the error generator and the drive motor needs to be DC-coupled, and have high gain and good stability. Typically a motor requires (under normal loading conditions) between 40 and 120 mA at 6 V, but the drive amplifier must be capable of supplying a much greater current than this to ensure quick acceleration from rest, and

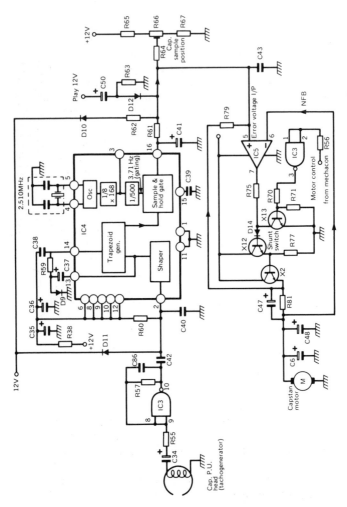

Figure 17.14. Capstan servo circuit with IC discriminator and discrete-transistor motor drive (Ferguson)

particularly to ensure rapid response in the motor to a change in drive from the servo system during normal phase and speed correction within the loop system. Implicit in this is the necessity for high 'loop gain' and we'll return to these characteristics later.

A discrete design of servo and motor-drive amplifier is the best instructional model (see *Figure 17.14*). The capstan servo is shown, and the rotational speed of the capstan is such that its tacho pulses appear at 3.71 Hz rate, much lower than the 25 Hz rate of the head drum pulses. In most other respects, the drum and capstan servos in this machine are similar. The upper part of the diagram is the phase detector, taking a positional feedback signal from the capstan tachogenerator and comparing it with a crystal reference (2.51 MHz, divided to 3.71 Hz) in the MSM5816 chip IC4. The error signal is developed on C39, and appears across C41 on pin 16 of the chip whence it is added, via R61, to a standing voltage of about 6.3 from the potential divider chain R65/66/67. The potentiometer R66 sets the DC operating conditions for the motor drive amplifier and is adjusted for correct 'free running' motor speed. This has the effect of restricting the servo-control effect to a narrow 'window' centred on 6 V or so, and ensures that the motor quickly runs up to speed from rest, aided during the play mode by the charging effect of C50 via D12.

Motor control voltage is fed into the non-inverting input of an operational amplifier IC5. The output from pin 7 of this chip passes to a Darlington pair X12 and X2, with the DC motor forming a load for X2. X13 forms a shunt switch operated by the mechanism control circuit. When pins 1 and 2 of the inverter IC3 go low, its pin 3 will go high, turning on X13 and grounding the base of X12. As a result X12, X2 and the motor will turn off.

A *discriminator-feedback* system is used in this motor drive amplifier, in which the reverse EMF (generator effect) of the motor is used to control the op-amp IC5. Motor current (proportional to load rather than speed) is monitored by 0.47 Ω resistor R81, and the voltage it develops is in effect applied across the inverting and non-inverting inputs of IC5, which then operates to balance motor voltage and control-input voltage. In addition, back EMF from the motor, appearing whenever its mechanical load decreases, is passed into IC5 as negative voltage feedback to its inverting pin 6, reducing motor drive to match the reduced loading on the motor.

Loop transient response and damping

The design of a servo system is critical if its performance is to be good, and VCR servo systems are very demanding in this respect. The performance of a servo is traditionally analysed in terms of a step input function, in which a step, or instantaneous change, in control voltage is applied and the output function (in our case, capstan or head drum phase position) analysed to see how closely it corresponds to the step input.

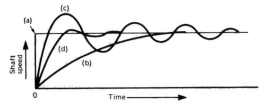

Figure 17.15. Loop transient responses compared. Waveform (a) is the step input; the other waveforms show possible reactions at the drive shaft

This is illustrated in *Figure 17.15*, where the step input is shown along with three possible responses at the servo output. Plainly it is not possible to achieve an instantaneous change in servo output where a mechanical system is involved – the inertia of the moving parts will cause an inevitable time lag before the new phase condition is established, and the designer's aim is to do this in the shortest possible time without spurious mechanical effects. Three factors are involved – damping, gain and time-constant, and

350

each of these are present in mechanical and electrical form. An ideal servo system has each of these factors optimised, and the electrical design of the motor drive amplifier takes into account the mechanical characteristics of the motor and shaft coupling systems, and inertia of the rotating parts. Electrical analogies of mechanical functions can be built in to a circuit in such a way that deficiencies are catered for – for instance, positive feedback via a capacitor will reinforce the gain of a motor drive amplifier only when the *rate of change* of drive amplifier output is high; the same effect can be realised by negative feedback via an integrator. This may be used to overcome the inertia effect of a heavy flywheel. Very often two time-constants are used to control the response time of the servo loop: one to quickly achieve correct speed (frequency) and a second to establish exact positioning (phase).

Let's now look at the servo responses illustrated in *Figure 17.15*. The step input is shown in curve (a). Curve (b) illustrates the effect of excessive damping, in which the output takes a long time to follow the input signal resulting in sluggish operation; curve (c) shows insufficient damping, causing the servo output to *hunt* or oscillate about the correct point for a period before settling down. Without any damping, continuous oscillation would take place! The ideal situation is shown in curve (d), that for *critical damping*. This is the one which designers normally aim for as offering the best compromise between response speed and hunting effect. A single slight overshoot and one small undershoot is followed by stable operation at the new level. This effect is easily demonstrated in most machines by making a sudden and well-defined adjustment of the tracking control to simulate a step input function to the servo control. The picture will be seen to move sideways in one or two damped cycles of oscillation before settling down.

There is great variation in the design of motor drive amplifiers, depending on the mechanical system and intended conditions of operation of the machine. Much depends on motor design, and the servo system we have described for the early Ferguson machine is one of the

351

simplest in use. Other contemporary designs were more sophisticated, and later developments like the portable machines now available, have more complex feedback arrangements in their servo systems. The development of brushless multipole DC drive motors with purpose-designed IC control systems has greatly improved servo performance in current VTR designs.

A digital servo

An alternative approach to servo comparator design is embodied in the digital system illustrated in greatly simplified form in *Figure 17.16*. Instead of the analogue ramp-and-sample

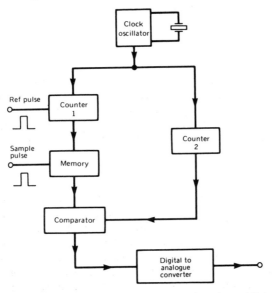

Figure 17.16. Simplified block diagram of one form of digital servo

arrangement, a counting system is used to derive an error voltage proportional to the time delay between reference and sample pulses. A clock oscillator runs at a fixed frequency to operate the system, and the arrival of a reference pulse *enables* counter 1, which starts to count clock pulses, feeding them into a memory. The memory is 'frozen' each time a sample pulse appears, and at that moment in time the stored count is proportional to elapsed time between the reference and sample pulses. We now need to compare this stored count with a reference, or time marker, and this appears in the form of counted-down clock pulses from counter 2. These are compared with the memory contents in a comparator whose output is converted to analogue form in a D-A convertor consisting of a bistable and charging capacitor. The error signal thus produced controls the drive motor in the normal way, and the circuit has the advantage of eliminating the presets and manual adjustments of the conventional sample and hold discriminator.

Still-frame requirements

Chapter 13 related how various machines solve the problem of noise-free freeze-frame reproduction by means of four-head drums or extra-wide replay heads to overcome the change in effective tape track angle between moving and stationary tape. When the tape transport is stopped at a random point, track scanning by the vision heads may be such that a mistracking bar is seen on the still frame.

It is necessary to position the tape so that this noise bar is 'shuttled' out of the picture to give satisfactory reproduction. This is carried out by the capstan servo which 'inches' the tape forward until the noise bar moves out of the picture into the field blanking period, when capstan drive ceases. The reference for the inching system is off-tape video, usually a sample of FM luminance from the DOC circuit, though certain Betamax-format machines use the sampled phase error of the replayed chrominance signal to achieve the same result. The presence of the noise bar in the field blanking period would tend to obliterate the field sync pulse during replay, and even if it were available, the half-line 'jump' in its timing between successive fields would cause picture-judder on still frame in machines where this system is used. For these reasons, an 'artificial' vertical sync pulse generator is used in still-frame mode, and its pulse timing is governed by a preset, adjusted for minimum vertical judder of frozen pictures.

An extension of the inching idea outlined above makes possible slow-motion replay in suitably equipped machines. Let's assume we need to replay at one-third normal speed. It will be necessary to position the tape accurately on the head drum to achieve good still-frame reproduction as described, and hold it in position for three sweeps of the video heads, i.e. three revolutions of the drum. After this the tape is quickly moved forward by the capstan to line up the next track accurately for the next three revolutions of the head-drum, and so on. The same process can be carried out with the capstan going backwards to give a slow-motion sequence in reverse. The precision required during these processes is of a high order and the circuits to carry out the operation use a mixture of analogue and digital techniques, often with specially designed ICs. A description is beyond the scope of this book!

Tension servo

During record or playback, it is important that the tension on the tape is held at a constant level, regardless of the amount of tape on each spool and the friction encountered by the tape on its journey through the deck. Insufficient tension in the tape during its wrap round the head drum leads to poor head-tape contact and results in poor signal transfer and noisy pictures. Too great a tension causes head and tape wear and tape stretching, and can result in poor compatibility with other machines due to 'stretch-distortion' of the magnetic pattern of the tape. When replaying a 'foreign' tape on a machine with excessive back tension, the registration between luminance and chrominance signals is lost, and the colour will be 'printed' to the left of the luminance image in the reproduced picture.

To prevent these problems, VTRs incorporate a tension servo to maintain a constant 'stretch factor' in the tape. In

354

simple machines this takes the form of a passive mechanical negative feedback system, in which a tension arm bears against the tape as it leaves the feed spool in the cassette. The arm is linked via a lever to a brake band wrapped around the supply spool turntable, so that low tension causes the brake band to bind against the supply turntable and vice-versa, maintaining reasonably constant tension in the tape. The arrangement is shown in *Figure 17.17*.

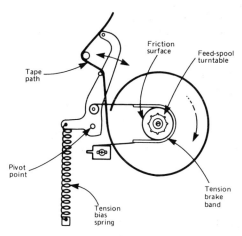

Figure 17.17. Simple brake-band tension regulator

In machines where each tape spool is driven from its own motor, a more precise control of tension is possible. Reverse current in the supply-spool motor sets the back-tension in the tape. Typically the tape emerging from the cassette passes over a spring-loaded feeler which carries a shutter; this is in the light path of an optocoupler (LED-phototransistor combination) so that conduction in the phototransistor is proportional to tape tension. The phototransistor current is amplified and used to control the current in the *supply* spool motor, which is arranged to run 'backwards' at this time. Because of the steady pull on the tape from the capstan, the supply motor will be pulled in the opposite

355

direction to which it is trying to run, and the motor current will then determine tape tension under the control of the optocoupler which it is trying to run, and the motor current will then determine tape tension under the control of the optocoupler sensor. At the beginning of a cassette, when all the tape is on the supply reel, the reverse current in the supply spool motor is of the order of 250 mA. As the tape is wound on to the

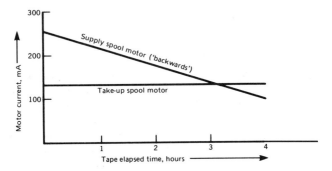

Figure 17.18. Electronic tension servo: Motor-control currents vary with the amount of tape on each spool

take-up reel this current drops to maintain correct tension, as illustrated in *Figure 17.18*, which also shows the constant forward current in the take-up spool motor, representing the latter's steady pull on the tape away from the capstan and pinch roller.

18

System control

Several times in this book, we have compared aspects of audio and video tape recording and the machines for doing so, and in each case we have seen how much more sophisticated is the video system. Simple audio tape recorders have only a single motor; no electronic circuit or motor design will make it rotate in opposite directions at the same time! All that is necessary, then, is a simple mechanical interlock to prevent more than one transport button being depressed at a time. In high-quality audio machines with fast spooling and more than one motor, a simple form of system control is provided. Once again, then, in the area of system control, the VTR has a multiplicity of requirements which do not exist in its audio counterpart. These arise from: the threading in and out of the tape from the cassette; the high rotational speed of the head drum; the fragility of the vision heads; the use of several drive motors whose efforts could conflict; the high speed and torque of the fast-wind and rewind processes; the need to control the machine by light-touch sensors or a remote keypad; and several minor factors.

Need for system control

The system control (we'll call it syscon for short) section of a VTR may be regarded as a 'policing' system for the mechanical functions of the machine. It has two basic interrelated functions, preventing user 'abuse' and programming the machine's mechanical operations, while protecting VTR and tape in the event of machine malfunction or tape faults. The syscon receives inputs from the user's mode keys (command functions) and also from monitoring points around the tape deck (protection functions). These are correlated in the syscon and so long as no conflict exists between command and deck or tape status, the mode required is permitted. If at any time during any mode, a relevant protection input is received (e.g. drum motor stops during play or tape-end sensor is activated during rewind), the syscon will invoke a new mode – except in microprocessor control systems, described later, this will be 'stop'! The syscon works in digital mode, and recognises on each input line only two states, on (1) and off (0). It presents its outputs to solenoids, motors etc. in the same way. As in previous chapters we will use 'discrete' circuits for illustration.

Syscon inputs

Let's examine the inputs to the syscon in turn. The command inputs are the user control keys – play, record, rewind, fast forward, fast rewind, pause, stop and, where provided, forward and reverse picture search. The tape counter also provides a command function (stop) if it reaches 0000 or a pre-programmed number ('go-to' facility) with its memory key depressed. In the VHS format, stop mode means un-thread then stop. The protection inputs are many and varied, and we'll look at each of them, and the effect they have.

End sensors. The tape-end sensors are activated by markers 150 mm or so from each end of the tape. Each format has a different form of marker and end sensor. For VHS a lamp shines through a transparent leader tape to activate a photo-cell, and the same applies to Video-8 format. All modern VTRs use an IR LED emitter. The Beta tape-end sensors work in a different way. An oscillator coil is held in position near the tape, on which the end marker consists of a metallic strip. When the two meet, magnetic loading of the coil takes place, the oscillator stalls and a tape-end signal is thus produced. In all cases, the tape-end signal invokes stop mode and then prevents any further tape transport in the 'unsafe' direction. VHS machines also incorporate a protection circuit which monitors the current in the tape lamp or LED; if an open-circuit here went undetected no tape-end signal could be given, and damage might result.

Slack sensor. The slack sensor may be mechanical (spring-loaded arm and magnet operating a reed switch) or optical (LED/photocell combination) and will invoke stop if a loop of tape should form along the tape path. This may be due to drive or clutch failure, or breakage of the tape.

Rotation detector. We have seen that the head drum generates tacho pulses as it rotates, and their cessation would mean that the head drum is stalled. This is monitored by the syscon, along with tacho pulses from the take-up spool carrier. If either fail, stop is entered after a short time. During freeze-frame or picture pause modes both tape spools will stop, so the spool rotation detector output is overridden in these modes. To prevent tape and head wear, a timer is started which enters stop after a few minutes' scanning time over a single tape track.

Cassette-in. All modes are inhibited when there is no cassette in the machine. A lever and microswitch detect the presence of a cassette, being deflected by its plastic envelope as the cassette carriage is driven home into the machine.

Recording tab. As with audio cassettes, video types have a removable safety tab to prevent accidental erasure or record-ing-over the programme it contains; where the tab has been removed, a lever and microswitch detect the fact, and signal the syscon to prevent record mode being entered.

Threading completion. Sometimes known as the after-load (AL) switch, this closes when the tape is fully threaded around the head drum. Until threading is complete, the syscon will inhibit capstan drive to prevent forward motion of the tape. *Mode-switches* give more comprehensive information.

Dew sensor. If the ambient temperature in which the VTR operates changes suddenly, usually due to bringing it in-doors from location work, there is a risk of vapour condensa-tion, particularly on the head drum. If the machine were allowed to operate in this condition the tape would stick to the head drum with disastrous results! To detect condensation, then, many machines are fitted with a dew sensor on the stationery part of the head-drum assembly; it consists of a pair of metallic fingers or a high-resistance 'element'. The presence of moisture will change the resistance of the sensor, and this is monitored by the syscon which acts to enter stop and illuminate a warning lamp. Some machines incorporate a low-wattage head heating element to drive off condensation. These features are important in portable equipment, and are present in many 'stationary' home VTRs.

Power interruption. When the machine is running in any normal mode, the syscon is in a 'latched' state, so that interruption of mains power will unlatch the circuit. The result, in many simple machines, is that stop will be entered on restoration of power.

As a corollary to this, if a piano-key control is depressed with a cassette in position but no mains power present, the syscon will release the key (stop mode) when power is applied. The exception is timer mode when keyed-in record instructions are held until the appointed time, then activated by a coincidence in the clock 'stored' and 'real' time.

Tape duration measurement. Late machines have a 'time-elapsed' indicator as a front-panel readout, and to operate this it is necessary to know what the cassette type is (2 hour, 3 hour etc.) and how much tape is left on each spool. Some cassettes have a code embossed into the cassette body to indicate running time, and this operates sensors in the VTR; some sophisticated machines of various formats compute spool speeds relative to each other against the fixed tape speed to arrive at a figure for cassette type in hours'-worth of tape. Once the tape running time is known, it remains to determine the amount of tape on each spool. This may be carried out via the tension servo system in some designs, or by an extension of the spool-speed-versus-tape-speed computation described above.

Syscon outputs

Piano-key operated machines used mechanical interlocks between keys, so that only one transport mode (forward, fast forward or fast rewind) can be entered at one time. This greatly simplifies syscon operation, and has the advantage that by linking the piano-key latching bar to a solenoid, stop mode (all keys up) can be invoked by momentarily energising it to release all keys. Even in a simple syscon like this, however, several other output lines are required, and we'll briefly examine these.

Motor stop. The capstan and drum motor can be enabled or disabled by on-off signals from the syscon. These usually operate transistor switches in the motor drive amplifiers – an example is X13 in *Figure 17.14.*

Pinch roller solenoid. This pulls the pinch roller into engagement with the capstan shaft to enable the latter to pull the tape through the machine. It will disengage during pause mode. In many Beta and V8 types, fast tape transport takes place with the tape laced up, and the pinch roller will be disengaged in fast-forward and rewind modes.

Reel brakes. In some VTR designs, tape reel brakes are operated mechanically from the user control keys. Other designs use magnetically applied brakes under the control of the syscon. During the threading phase the spools need to be unbraked to enable a loop of tape to be drawn out of the cassette. During unthreading, spool drive needs to be applied to take up the tape loop, pulling in the slack as it becomes available.

Indication lights. The syscon drives panel-mounted LEDs to indicate machine and tape status. The number and sophistication of these depends very much on make, model (and price!) of the VTR.

Programme muting. To prevent unlocked pictures and mistracking effects from being seen as the machine runs up to speed (*lock-up time*) the VTR's video – and sometimes audio – sections are muted, either for a fixed time-period after play mode is entered, or until the servo systems are locked up. In the latter case, the mute will operate in the absence of off-tape tracking pulses (e.g. blank or faulty tape), and in the event of servo malfunction. This automatic mute arrangement is common in Betamax machines and in later models of other formats. During 'trick-speed' operation, such as pause, shuttle search and slow or reverse motion, the servos will not be operating normally, and the syscon must override the auto-mute circuit.

Syscon circuit

We have seen that all inputs and outputs from the syscon are in binary form, or logic levels 0 or 1, and that the syscon function is one of decision making. This means that a syscon can be made entirely of logic gates, invertors and counters, though interfacing circuits are required to enable their outputs to drive solenoids, motors, brakes and so on.

First-generation VTRs used transistors and diodes in their syscon circuits, and the requirements lent themselves to the use of simple digital IC arrays such as 74-series TTL chips. The next logical (!) step was to integrate all logic functions into a single chip containing the necessary gates and counters for a particular machine's requirements, and this idea was adopted in many piano-key operated machines in all formats. A syscon of this sort is illustrated in *Figure 18.1*, together with its interfacing circuits. The internal workings of the chip are fairly obvious from the diagram, but it is instructive to look at its peripheral circuits and interfaces, as they give a useful insight into the overall operation of the machine and an introduction to more sophisticated syscon designs in later machines. We'll examine circuit operation from the point of view of each IC pin.

Pin 1: 'Operate' in. Goes high in all modes except stop, and primes the chip for further commands.

Pin 2: Rewind in. High when rewind is selected.

Pin 3: Play in. High when play key selected.

Pin 4: Pause in 2. Fed from X11; high during pause mode to initiate 15-minute pause counter within the chip. At the end of the count, stop mode will be entered.

Pin 5: Pause output. High during pause; feeds pin 4 and pause rail out via X11.

Pin 6: Motor control. High during drive modes; this is turned to a low in inverters IC3 to switch off motor control transistors X12 and X13, thus enabling drive motors.

362

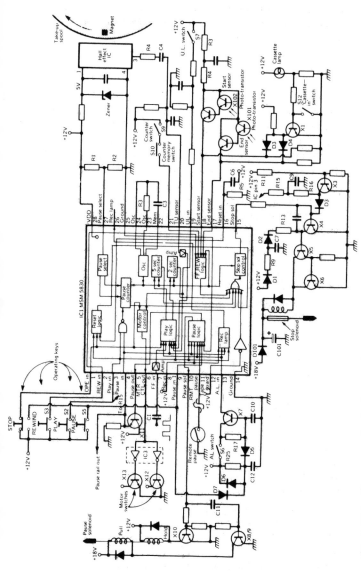

Figure 18.1. The syscon circuit of a piano-key operated VTR. The purpose-designed IC is surrounded by interfacing circuits for sensors and 'operators' such as solenoids and motor switches (JVC)

363

Pin 7: Flip-flop in. Drum rotation detector input, in the form of head drum tacho flip-flop pulses. Their absence will trigger stop mode after 5 seconds, timed by the stop counter within the chip.

Pin 8: Record in. Not used in this machine, and held permanently at 12 V (high).

Pin 9: Pause in 1. High when pause key down, and in stop mode. Releases pause solenoid to stop tape transport.

Pin 10: Pause solenoid output. Goes high during pause mode; this turns on X10, and also X8 and X9 via C11. X9 draws a heavy current through the upper winding of the pause solenoid from the 18 V supply and the armature pulls in. Within a second or two C11 has charged and X8/9 turn off, but the armature is held in by the much smaller sustained current from the 12 V line through the lower ('hold') solenoid winding and X10.

Pin 11: Remote pause in. Takes command from a camera or wired remote pause control via a rear-mounted jack. Normally grounded, high to pause.

Pin 12: Tape-guard input. Not used in this machine, and held permanently at 12 V (high).

Pin 13: After-load in. When the tape is fully threaded-up, S6 closes, and after a short pause during which C10 charges through R17, X7 switches on, putting a high on chip pin 13, whereupon the pause solenoid engages and tape transport begins. During threading, the motors and servos are operating, so that they are locked up by the time tape transport starts. If no AL signal is received within five seconds of requesting play, the syscon will enter stop.

This machine has facilities for serial recording, which means that it can be timer-programmed to record a series of TV programmes on consecutive days. During this mode, the tape may be already threaded-up when unattended recording begins, and unless precautions are taken, at the appointed time the motors will start from rest, run up to speed and 'lock in' while recording is taking place. This would result in a very ragged start to the recording, a situation that is prevented by the circuit connected to the base of X7. Under the circumstances described, the AL switch

364

will be closed when recording begins, initiated by the appearance of +12 V at S6. C10 cannot now charge quickly because its charging current is diverted by D5 into the 'empty' C12, so that the charge time for C10 is now tied to the much longer time constant C12/R25. X7's switch-on is thus delayed to enable servo locking to be completed before the pause solenoid is energised to initiate tape transport. Should power be interrupted during operation, D6 rapidly discharges C12 into the decaying +12 V line, ensuring the long time-constant when power is restored. D7 maintains full charge on C12 during modes other than timed serial recording; this ensures that D5 is reverse-biased isolating the long time-constant circuit from X7 base.

Pin 14: IC ground for operating potential.

Pin 15: Stop solenoid out. Normally high, this goes low during stop mode to turn off X4. When X4 goes off, current flows via D1, R9, D2 and R13 to turn on X5 and X6 and energise the stop solenoid. Because the solenoid is automatically de-energised (by the cessation of 'ope in' potential on chip pin 1) when stop is achieved, the stop solenoid has only a single power-winding working from the +18 V supply. This contrasts with the separate 'pull-in' and 'hold' windings of the pause solenoid, which is required to be energised for long periods.

If, because of operator abuse, a function key is held down during 'protection' conditions, the syscon will hold off motor drive (via pin 6) and pulse the stop solenoid at two-second intervals until the key is released; this is the function of the 2 second timer within the chip.

A 'conditional' stop function is provided by the circuit centred on X3, and it governs the stop solenoid operation during mains power failure. During play or record mode, S2 ('play' key) is closed and X3 collector potential low due to its base 'turn-on' bias via R11. D3 is thus reverse-biased and has no effect on circuit operation. If the mains supply fails, the +12 V line collapses, X3 base bias disappears and it turns off. D3 now starts to conduct and the charge on C9 flows through R16 to maintain base current in X4, holding off X5, X6 and the stop solenoid until all power lines have decayed to zero.

Thus the keys remain down, and normal operation is restored when the mains returns. If power failure should occur during fast-wind or rewind, however, S2 will be open and C9 discharged. X4 will thus be allowed to turn off and the stored energy in C7 (isolated from the decaying +12 V line by D1, now reverse-biased) will provide base current for X5 so that the stop solenoid will be energised. Solenoid energy comes from the large capacitor C101 (isolated from the decaying +18 V line by D101). Enough energy is stored in C101 to pulse the stop solenoid once at the instant of mains failure.

Pin 16: Reset in. At switch-on, all functions in the syscon chip must be reset. This is done by a positive pulse from the differentiator C6/R5 operating from +12 V.

Pin 17: End sensor. A high on this pin will initiate stop mode, and while it remains high the syscon will only permit rewind mode. The cassette lamp shining through clear 'end leader' tape will illuminate end-sensor photo-transistors X101, causing them to conduct via R3 and R4.

Pin 18: Start sensor. As for pin 17, but the high signalled by start-sensor photo-transistors X102 will cause the syscon to inhibit rewind mode.

X1 also operates on chip pins 17 and 18. Normally on, it holds D3 and D4 off by grounding their anodes. If the cassette lamp should fail, X3 will be robbed of base current and will turn off, raising anode potentials on D3 and D4 to put a high on chip pins 17 and 18, invoking stop mode and preventing motor drive. The same action occurs in the absence of a cassette, when S12 is closed, grounding X1 base.

Pin 19: Unload in. Operated by unload microswitch S7, which maintains a high on chip pin 19 until tape unthreading is complete, whereupon the syscon switches off drive motors via its pin 6. A second unload microswitch (not shown in *Figure 18.1*) operates to hold in the 'timer-on' relay until unthreading is complete so that the machine goes to stop at the end of a single timed (unattended) recording.

Pin 20: Take-up sensor in. To detect rotation of the take-up spool a *Hall IC* is used. The device is mounted close to a rotating magnet driven from the take-up spool and responds to the nearby alternating magnetic field by an internal 'deflection' process explained in *Figure 18.2*. The AC output from the Hall IC is fed via R4 and C4 to chip pin 20 where it is monitored; its cessation indicates that the take-up spool has stopped, and after five seconds delay the syscon will enter stop.

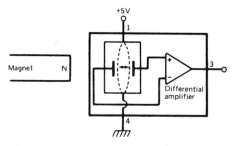

Figure 18.2. Internal operation of the Hall-effect IC. The flow of electrons through the Hall element on the left is, in the absence of a magnetic field, down the centre. An alternating magnetic field will deflect the electron stream to right and left, rather like an electron beam in a TV picture tube. The alternating charges picked up by the side electrodes are amplified and passed out on pin 3 of the chip

Pin 21: Memory in. Memory in this case means 'tape counter at 0000'. A microswitch S9 within the counter will close at this point, and provided the operator's memory switch S10 is 'on', a high is presented to chip pin 21; the syscon will enter stop.

Pins 22, 23 and 24: Clock timing. Within the syscon chip is a 20 Hz oscillator for use with the 2 second (stop-cycling), 5 second (stop protection delay) and 15 minute (pause timer) count-down dividers. The 50 ms period of this clock oscillator is set by R3 and C3 connected to chip pins 22, 23 and 24.

Pin 25: Pause logic in. Unused in this machine, and grounded.

Pin 26: Record lamp out. Not used in this machine.

Pin 27: Pause select. The pause timer delay is governed by the potential applied to this pin. Current video tapes can withstand scanning of a single track for several minutes, and the potential divider R1/R2 sets 6 V on chip pin 27 to give

367

approximately this delay before invoking stop. The range 0–12 V at pin 27 offers delays from 7 to about 30 minutes.

Pin 28: VDD in. This is the operating potential for the logic circuits within the chip itself, +12 V.

Simple syscon summary

The system described and illustrated above is effective, relatively simple and virtually foolproof. Inherent in the design, however, are several limitations: remote control is limited to a single *wired* pause function; machine operation is necessarily via 'clunky' piano keys which operate many deck functions by direct mechanical means; and unattended recording is limited to a single programme during the three days (or one week, depending on the clock capabilities) following clock setting, or serial recording from one TV channel at a fixed time of each day. The system has the advantage of low cost and economy of drive motors; as mentioned in Chapter 17, some machines with syscons similar to the one we have described used but a single motor for all mechanical functions.

Touch button deck control

Perhaps the greatest objection to mechanically operated VTRs is the heavily-loaded piano-key operating controls, and these may be compared with the mechanical tuner buttons of early TV receivers. In TV design the arrival of electronic (*varicap*) tuners paved the way for the provision of electronic station selection via light-touch (microswitch) buttons, or touch sensors in which the mere presence of a finger on a sensitive surface triggered on an electronic latching circuit to select the required channel. The next generation of home VTRs, then, offered light-touch selection of deck functions, but at a cost of much greater complexity in the syscon department, on two counts: the loss of the mechanical interlock facility, and the necessity to provide motors and

368

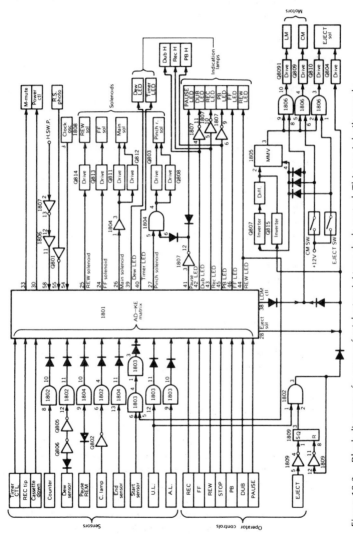

Figure 18.3. Block diagram of a syscon for 'electronic' tape deck control. This one is built round a microprocessor IC.(Sharp)

369

solenoids to do the mechanical chores previously performed by the operator via levers connected to the piano keys.

The loss of mechanical interlock between selectors means that the operator can enter 'illegitimate' commands such as fast-forward during the rewind process, or play and stop at the same time. This requires the syscon to 'think harder', as it were, assigning priorities and storing user instructions until it has achieved a deck status which permits the newly-requested mode. The requirement for the deck to do its own housework, so to speak, multiplied the number of syscon outputs, with spool brakes, spool drive motors, eject solenoid, threading motor and comprehensive indicator LED systems having to be provided and very precisely controlled.

While this could be achieved by an extension of the standard logic techniques of the MSM5830 chip described earlier, a great deal of circuitry would be required, so an alternative technique is used. This is the microprocessor control system and a syscon using this type of chip is shown in block diagram form in *Figure 18.3*. Inputs, both command and protection, enter the chip on the left, and outputs leave on the right. A great deal of interfacing is required between the syscon chip and the functional parts of the machine, and IC1801 in *Figure 18.3* is supported by eight further chips and a multiplicity of transistors and diodes. The control keyboard has grown up from a series of lever-operated microswitches to a matrix system like that in a pocket calculator – pressing a button links two input pins on the syscon chip and the command required is interpreted within the chip.

Microprocessor control of full function machines

The basic microprocessor control system in *Figure 18.3* caters only for mechanical control of the deck, and may be regarded as an 'electronic' version of the simple syscon previously described. For complete versatility in VTR control and facilities, an extension of the syscon is required to cater for further functions. Complete cordless remote control of all deck modes, including trick-speed effects, was introduced.

Multi-event timers, in which different broadcast channels could be programmed for recording at various future times and dates, became available. Front-loading machines appeared with the means of mechanically transporting the cassette into its carrier, then lowering the carrier itself into position on the deck. 'Clever' editing of tapes became possible.

Machines incorporating these features need a very sophisticated syscon akin to a small computer to correlate, organise and carry out all necessary control and protection functions, and our earlier analogy of system-policing now has an additional comprehensive 'housekeeping' function. This is an ideal role for the microprocessor, and these devices are now commonplace in domestic VTRs, not only in syscon circuits but increasingly in such areas as servos and clock/display electronics.

Basic microprocessor operation

A microprocessor is an LSI (large-scale integration) integrated circuit. It works in quite a different way to such logic chips as the 74 series and the MSM 5830 type described earlier, in that it functions in similar manner to the CPU (central processing unit) of a computer. In normal computer applications it is supported by one or more separate memory chips, in which data is stored and retrieved by the CPU as required.

The sorts of microprocessors found in VTR machines, however, need very little memory capacity, and it is possible to design a complete simple computer system into a single package, perhaps better termed as in 'integrated microcomputer'. It contains all the elements of a small general-purpose computer on a single chip. Thus (*Figure 18.4*) we find a CPU (central processing unit), ROM (read-only memory) RAM (random-access memory) CLOCK, and I/O (input/output) PORTS. The CPU is the heart of the device, controlling the interchange of information between the various sections of the chip and the I/O ports. The ROM contains control programmes covering all combinations of

371

input signals to the system, and these instruct the CPU when the latter calls for them in turn. The RAM is used as a temporary 'storeroom' for data which needs to be held pending the completion of mechanical or data-processing operations. Input and output ports accept requests and

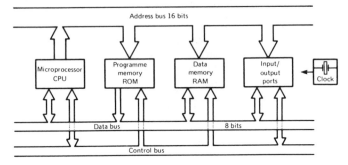

Figure 18.4. Internal block diagram of a microprocessor IC

dispense commands to and from the microprocessor. The clock is a steady source of timing pulses which 'steps' data sequentially through and around the microprocessor. Address, data and control bus systems are the internal highways for data interchange between the sections of the chip; data is loaded into these on a sequential basis, preceded by an electronic 'label' to indicate its routing and destination.

Under the control of its internal and pre-programmed 'action memory' the micro can cope with any combination of circumstances, and come up with the most logical answer. Taking our earlier example of the user keying in fast-forward during a rewind operation, the micro-syscon will enter stop, wait for that to happen, then programme the deck for fast-forward operation. In other designs, fast-forward will be entered immediately, with reel motor drive and braking controlled in such a way that the changeover is effected in the shortest possible time consistent with not stretching or 'looping' the tape. If more memory or processing capacity

372

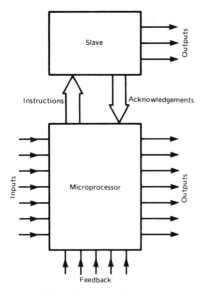

Figure 18.5. The versatility of the microprocessor can be increased by the addition of a 'slave' chip

than can be accommodated in one microprocessor is required, then an additional slave chip can be tacked on, in the manner of *Figure 18.5*.

Multiplexing

In the sorts of machine we've described, it's not hard to imagine that the number of command and feedback inputs, and control outputs from the syscon chip will run to over 60. It is impractical to provide more than 40 or so pins on an IC, so that a time-sharing system, known as multiplexing, is used on input and output ports. This is carried out by means of expander chips. These work on a sequential basis, switching several data lines to and from the micro ports in turn when requested by the control or 'chip-enable' line from the

373

micro's CPU. During the 'waiting' period, the expander chip stores data in its own internal memory.

Key scanning

A further economy in data lines is achieved by the process of sequential scanning of the control keyboard. The keyboard is wired as shown in *Figure 18.6,* and is fed at the top with 'strobed' scanning pulses of perhaps 1 ms duration in sequence, so that SP1 is energised for 1 ms, then SP2, SP3 and SP4.

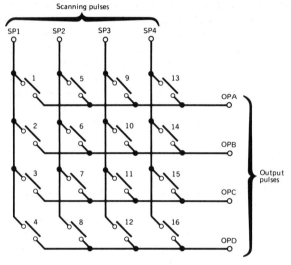

Figure 18.6. Key-scanning matrix for economy in wiring and IC pins. Operation is explained in the text

The output lines are monitored by the micro for returning pulses, so that if line OPB 'lights up' during SP3, button 10 has been pressed; if line OPD comes on during SP4, button 16 has been selected, and so on. The returning pulse from an OP line is decoded within the micro and stored in its memory as a command to be processed.

374

This key-scanning system used a total of eight lines to handle 16 keyboard commands. The number of keyboard lines can be further reduced by more complex scanning systems. Some VTR designs, notably camcorders, have a scanning system embracing not only control keys but such things as deck sensors.

Remote control

For cordless remote control of VCR functions, an infra-red link is used between the VTR and a hand-held commander unit. The latter consists of a keyboard linked to a digital encoder chip which prefaces each command by a *framing code* (which identifies the commander and primes the decoder), and then sends a string of binary pulses called a word, each unique to the mode requested. With only a single infra-red link, the data is necessarily sent in serial form. On receipt in the VTR, a *shift register* converts the serial data to parallel form ('static' data on separate lines) for scrutiny by the keyboard scanning system of the microprocessor.

Micro interfacing

The number and complexity of interfacing devices depend on the degree of dedication of the microprocessor. Late machines use purpose-designed micros which are relatively simple in their interfacing requirements; up-market Sony machines, for example, have separate 'pull-in' and 'hold' micro outputs for solenoids; even 'analogue' outputs can be produced by a variable mark/space ratio pulse. A general block diagram for microprocessor control of a VTR is given in *Figure 18.7*.

Summary

In this chapter we have examined the role and operation of the VTR's system control section, and explored the interfaces

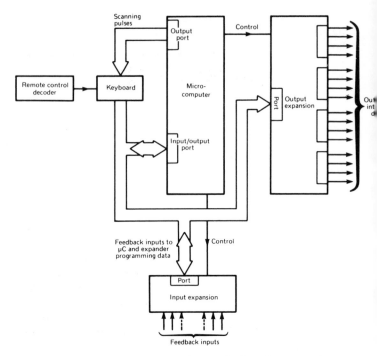

Figure 18.7. Microprocessor-based syscon for 'housekeeping' within a VTR

by which it communicates with the 'outside world' of keyboard, tape deck and signal circuits. The evolution of syscon design has been traced from the simple logic and mechanical deck control of the early 'clunk and twang' machines to the computer-based technology of modern full-specification models, though it has not been possible to treat the latter in any great depth.

19

The complete VTR

The circuits and processes discussed earlier in this book form the basics of a home VTR and are, for the most part, unique (in the domestic environment) to videotape machines. In this chapter we will deal with the peripheral circuits, i.e. those which take no part in the basic video recording process but are essential to 'service' and power the machine and provide operating convenience and flexibility. All the circuits to be discussed have counterparts in other domestic equipment such as clock-radios, TV games, audio tape recorders and TV receivers. *Figure 19.1* shows a block diagram of a VTR.

Mixer/booster

To avoid regular plugging and unplugging operations, the VTR is permanently connected 'in series' with the aerial lead to the TV receiver. When the machine is off, or recording a programme other than the one being viewed, it is important that normal TV reception is not affected by the machine's presence, so a *loop through* facility is provided in the aerial booster, a small RF amplifier which is permanently powered. Its modest gain cancels the losses incurred in the extra RF plugs, sockets and internal splitting of the RF signal within the VTR.

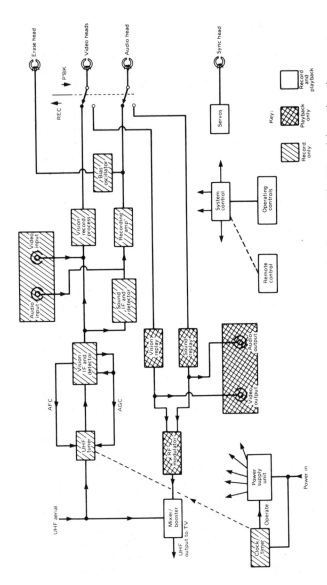

Figure 19.1. Block diagram of a VTR, with emphasis on the 'peripheral' circuits. The RF modulator, though marked for playback only, is often used during monitoring of recordings

UHF tuner and IF amplifiers

Most recordings made on a home VTR come via broadcast transmissions, so the machine needs a tuner and receiver built in to select and demodulate broadcast programmes. At the time of writing, all UK programmes are radiated in the UHF band, and the tuner and IF arrangements are identical to those provided in contemporary TV sets. An effective AGC circuit is provided to ensure a constant signal level to the recording section, and AFC feedback maintains correct RF tuning. An increasing trend is the provision of self-seek and memory tuning, in which the tuning potentiometer bank is replaced by two or three push-buttons and an IC 'brainbox'.

Sound

An intercarrier IF amplifier and detector follows the vision detector to provide a sound signal for recording on the tape. After the detector, the sound recording processes follow audio cassette practice, with the audio signal being added to an HF bias source before recording on a longitudinal track on one edge of the tape (as shown in *Figure 13.5*). The slow speed of the tape and narrow audio tape track do not make for ideal recording conditions, and noise reduction systems are commonly used to improve performance. Foremost among these is the Dolby system, which involves a form of non-linear and frequency-dependent pre-emphasis during the recording process, and complementary de-emphasis in replay.

In 1982 a new feature was offered in domestic VTRs, that of stereo sound capability. The audio track is split into two with an intervening guard band, shown in *Figure 19.2*. For tape interchangeability and electrical compatibility, both tracks are used in 'parallel' for mono sound recording. This longitudinal stereo system was soon rendered obsolete by the hi-fi sound systems described in Chapter 16. The introduction of the Nicam stereo transmitting system permits some programmes to be recorded and replayed in stereo; many pre-recorded cassettes

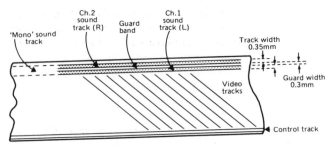

Figure 19.2. The sound track split into two for stereo sound (VHS system)

offered for sale or rent also incorporate good stereo sound tracks.

The concept of stereo-with-TV, be it from broadcast or video-tape, is a difficult one. The whole reason-for-being of a stereo setup is to recreate a wide, vibrant and 'living' sound stage in the listening area. Even with a very large screen by current standards the picture size (and position, very often) has little correlation with the sound field in most domestic situations. Stereo TV sets are becoming common, and while they are much better than early monaural models, the loudspeaker spacing is necessarily closer, drawing in and tightening up the sound stage to match the small picture 'window'. This is not ideal for concerts and recitals, whose main feature is the sound itself! Ideally, link a hi-fi VTR to hi-fi amplifiers and loudspeakers.

Bias oscillator

To provide a suitable AC recording bias for the 'longitudinal' audio signal, and to generate an erasing signal to wipe out video tracks, a power oscillator is used. Usually consisting of a discrete transistor oscillator built round a feedback/driver transformer, it operates at about 60 kHz, and is used only in record mode. Besides the audio head it powers a full-width erase head to wipe all signals, sound, picture, and control track, off the tape on its journey towards the head drum during recording. As described in earlier chapters, no re-cording bias signal is required for the video heads as this function is performed for luminance and chrominance by the constant-level FM luminance recording signal.

Clock/timer

Depending on vintage and price, VTRs vary greatly in their timing and programming facilities. All machines have a front-panel clock display, analogue (face and hands!) in early Philips machines, fluorescent digital in mains-powered static machines and LCD (reflective) digital in battery-operated portables to conserve power. The clock is arranged to show real time as a permanent display, counted down from a local accurate reference – usually a quartz crystal, sometimes 50 Hz mains frequency. A digital memory in the clock/timer chip is capable of storing the clock display data for a pre-set time, and this stored data is continually compared with the real-time display. When the two correspond exactly, a coinci-dence-detector opens a gate to switch on a power relay in the power supply unit. The VTR now comes to life in whatever mode of operation has been set up previously, in just the same way as a clock-radio, or clock-cassette/radio.

In the first machines, that was the limit of operation of the clock/timer, except that in some models the clock display doubled as tape counter indicator when this function is selected by the user. Late designs of VTR have comprehen-sive timer facilities, typically offering a choice of eight pre-programmed recording 'events' of any desired length, on any channel over 14 consecutive unattended days. This is achieved by an extension of the above technique, with the clock detector now looking for a day-coincidence as well as time-display coincidence, then gating on the broadcast chan-nel held in its memory as well as record mode for the tape deck. These are then held on for the requested time. Such a sophisticated timer circuit is often based on a micro-processor (see Chapter 18) which is capable of informing the operator if he keys in impossible requests like 'start at

1930 hrs, stop at 1900 hrs, or overlapped recording instructions on different channels!

RF modulator

When the machine is playing back, a TV signal at baseband is produced, and this is difficult to apply to an older type TV receiver, whose only signal input facility is via its aerial socket. To cater for this, an RF modulator is provided within the VTR, working as a tiny TV transmitter. Baseband audio and video signals are applied to its input for modulation on to a UHF carrier, whose frequency is chosen to fall into a gap in the broadcast spectrum around channel 36. A rear-access preset control will swing this frequency by about ±4 channels to avoid beat effects with other RF signals in the vicinity.

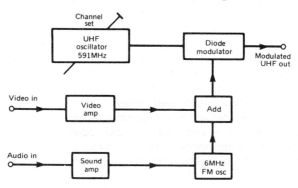

Figure 19.3. Basics of UHF RF modulator

Modulator characteristics are according to CCIR system I, the UK standard, and a diagram of the modulator is given in *Figure 19.3*. Output level is set to be 1–3 mV (the optimum for a TV receiver), and this modulated RF carrier signal is added to the booster output for application to the VTR's RF output socket.

During record, regardless of the signal source which may be off-air, TV camera, cable, another VTR or whatever, the

signal being recorded on tape is applied to the RF modulator
so that it may be monitored on the TV set if desired. This is
called the E-E mode (Electronics to Electronics) to distinguish
it from off-tape playback mode. The E-E signal is taken off
from the record electronics as late as possible, and several of
the record and playback circuits (such as the luminance/
chrominance adding stage) are usually included in the loop.

Set channel facility

During initial setting up, when neither the TV nor the VTR are
correctly tuned, it is difficult to establish the correct tuning
point for the TV when adjusting it to the output channel of
the VTR's RF modulator, unless a pre-recorded tape is available to provide a playback signal. To assist with this, most
machines have a simple video pattern generator incorporated, which modulates the RF carrier with a 'marker' to
enable the TV to be tuned. It is often part of the multi-
position colour-killer switch mentioned in Chapter 15, and is
mounted on the rear of the VTR.

Audio/video in/out connectors

Many of the signals required to be recorded are are baseband
for video and audio, typically coming from a TV camera or
from another VTR for copying/editing. Sockets are provided in
the form of DIN or phono for sound, DIN, BNC or phono for
vision, and increasingly a 21-pin SCART socket for both. Specifically for cameras, which will have pause, power, audio and
video lines, a multiway socket is usually provided, but regrettably there is little standardisation in these, and each manufacturer's VTR will generally only interface with the same maker's
camera.

Baseband outputs are also provided in similar connectors
to those mentioned above, so that connection to audio
systems, TV *monitors* and other equipment is possible,

including, where applicable, the use of the electronic view-finder tube in the camera as a playback monitor. The standard for baseband video signals is universal at 1 V peak-to-peak (sync tip to peak white) though sadly no similar common standard exists for audio signals.

Record-playback switching

In almost every department of the VTR, from sound to servos, from tuner to syscon, function switching is required between record and playback modes. A multi-way slide switch to achieve this would be physically very clumsy and difficult to arrange, though some early piano-key operated machines had one or two slide switches on the audio-video recording panel, operated by levers from the function keys. In these and later 'switchless' models, the bulk of the record/playback switching is carried out by electronic switches in the form of diodes, transistors and ICs, toggled by supply lines which appear during the relevant mode. Thus we find '+12 V record' or '+9 V play' lines distributed to all operational blocks of the VTR. While the source of these lines is the PSU, they are enabled by the user's operating keys, usually via the syscon.

Power supplies

A home VTR consists, so far as its electrical circuits are concerned, of a lot of transistor/IC electronic circuits, a few relays and solenoids and three to five low-voltage DC motors. There are a few exceptions to this rule, such as the large AC motor in some VCR and Betamax models, and the piezo head-actuators in old V2000-format machines, which require direct mains and ±180 V DC lines respectively. Most VCR circuitry operates at a supply-line voltage of between +9 and +20 volts (typically +12 V), with some critical lines requiring close stabilisation and decoupling. Solenoids call more for power than precision in their supply lines, so

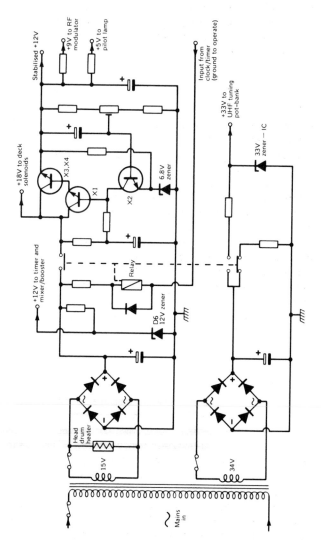

Figure 19.4. Simple power supply as used in early VTR machines (JVC)

unstabilised PSU outputs at low impedance are provided for them. Mains-operated PSU's, then, usually start with a mains transformer having several low-voltage secondary windings, each feeding a bridge rectifier followed by smoothing networks and voltage stabilisers. A typical PSU design is shown in *Figure 19.4*. The design is conventional and straightforward. X3/X4 form a DC series stabiliser in a control loop circuit embracing X1 and X2. The relay closes when the timer 'perks-up' and enables the main +12 V and +18 V lines. An 'ever-12 V' line for clock/timer and mix/booster is furnished by zener D6.

In industrial and TV applications, switch-mode power-supply units (SMPSU) have been common for many years; they have the advantages of higher efficiency and lower heat-dissipation over conventional circuits. The first example of SMPSU design for domestic VCRs made its appearance in second-generation Sony machines, and a circuit for the PSU of their model C7 appears in *Figure 19.5*. The purpose-designed IC generates a squarewave drive for the power switching transistors Q1 and Q2. A square wave is developed in the primary winding of T102, whose secondaries feed a low voltage supply to a full-wave rectifier and smoothing system. Stabilisation is achieved by a 'potted-down' sample of the 12 v output which operates to vary the mark-space ratio of the square wave generated within the chip. A soft-start feature is provided by C211 and Q3, which is fed at its base with an 'enable' line from the clock/timer – Q3 thus takes the place of the relay in *Figure 19.4*. Over-current and over-voltage protection is incorporated in the chip.

The switching transistors Q1 and Q2 are, at a given instant, either hard on or turned off, and in neither state do they dissipate much energy. This, and the relatively high frequency at which the circuit operates, leads to high efficiency, reliability and good regulation of output voltage.

Some VTR circuits require negative supply lines at low energy, a typical example being the fluorescent clock display, which calls for an operating potential of about −20 V and a filament supply of 2.5 V AC. These are inconvenient to generate and distribute from the main power supply, and a

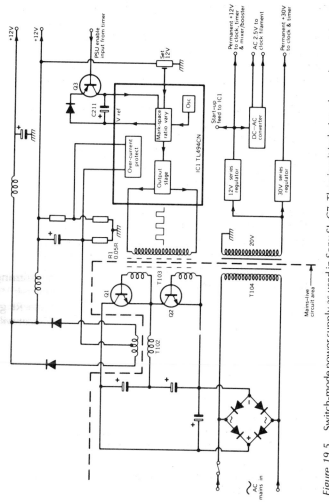

Figure 19.5. Switch-mode power supply as used in Sony SL-C7. The circuit is designed round a special chip, TL494CN, and offers high efficiency and close stabilisation of the two upper 12 v lines. Mains isolation takes place in transformers T102, T103 and T104. R1 is a current-monitoring resistor

DC-DC converter is often employed, conveniently sited at the point where its output is needed, and operating from the common VTR +10 V or +12 V supply. It is a mini-switch-mode power supply, working to a similar principle as the large version in *Figure 19.5*, but in a package the size of a postage stamp!

Clock back-up

When the VTR is operating or in standby mode, power failure will stop its clock; this is irritating in real time, and disastrous where the machine has been primed for unattended recording. To prevent this, a small rechargeable on-board battery, usually Ni-Cad type, is provided to maintain power to the clock oscillator and counters, so that when power is restored the display lights up to the correct time. Where *volatile* memories are used, the back-up battery must also power these to prevent loss of stored data.

Increasingly, *non-volatile* memory chips are appearing in VTR equipment, and these are able to retain stored data relevant to TV channel selection, deck status and timer instructions for a long period with no power supply of any sort.

20

Care, operation and maintenance of VTRs

A videocassette recorder is a complex ensemble of mechanical, electrical and electronic parts working to very high precision. As with any such equipment, careful handling and regular routine maintenance are necessary to maintain performance throughout the life of the machine.

Operation of the domestic VTR is relatively simple with the exception of some clock-programming procedures, when the instruction book must be slavishly followed! Much thought has been given by designers to ease of operation, and the cassette system and syscon between them make the machine virtually foolproof. The VTR should be installed on a strong, level surface in a position where the ambient temperature is reasonably constant and the atmosphere clean – the 'chimney corner' is definitely out on both these counts! It is recommended to buy a dust cover – regular use of this will prevent ingress of atmospheric dust and soot particles, so prolonging the machine's life. In high ambient light, and particularly direct sunlight, the usable range of any infra-red remote control system will be greatly reduced, and this should be borne in mind during installation.

If the VTR is in close proximity to the TV receiver or monitor, mutual interference effects can take place due to direct radiation of signals between them. This will lead to spurious effects ranging from patterning and striations to complete or partial loss of colour. While the specific cause of the effect may be obscure, physical separation of the TV and VTR will prove the point, and these symptoms can often be cured by installing a metal or foil screen between the two equipments, and earthing it if necessary.

Tape care

Considering its nature, video tape is surprisingly uncritical in storage and handling, largely due to the protective cassette housing. Cassettes should not be placed near magnetic field sources such as TV sets, large loudspeakers or transformers; they should be stored standing upright to prevent sag or cinching, preferably in an even temperature away from dust.

When a tape is used for 'time-shift' or similar purposes, the temptation is to rewind it to the beginning before each recording so as to have maximum recording time available. This practice leads to disproportional wear on the first part of the tape, resulting eventually in noisy recordings with lots of drop-out. Very often a three-hour tape is used for general purposes and the first 30 minutes-worth can become worn before the tail-end of the tape has seen any action at all! It is far better to work through all the tape, using the counter memory, before starting again at the beginning.

Head-cleaning tapes

The use of proprietory cleaning tape cassettes for domestic VTR head-cleaning is something of an emotive issue, with contradictory claims being made for their effectiveness, and much more important, the possibility of their causing damage to the video heads, by their makers on the one hand, and VTR manufacturers and the service trade on the other. If a head-cleaning tape must be used, the soundest advice we can give is to use a type marketed or recommended by the VTR manufacturer. Leaving the head-damage claims aside, however, there is no doubt that hand-cleaning of the heads (and the rest of the deck) is far more effective than any cleaning tape, although the former course is not open to the average VTR user, except by arranging service from a specialist service organisation.

Deck servicing

It must be emphasised here that any attempt to clean tape heads or adjust deck components by a person without the necessary knowledge and equipment is almost bound to end in disaster! One half-turn on a tape-guide adjustment screw can destroy the machine's compatibility, and the lightest of cleaning strokes in the wrong direction across a video head will destroy it; the repair bill could run to three figures. VTRs are not like the car or the central heating system!

Deck servicing resolves into three main areas, those of cleaning heads, tape path and deck surfaces; adjustment of tape guides and tensions to ensure compatibility; and attention to secondary mechanical parts such as brakes, clutches, drive wheels and operating levers and sliders. We'll look at each of these aspects in turn.

Cleaning

In Chapter 13 we studied the requirements of compatibility and saw that this calls for great accuracy in the laying-down and reading-off of video tape tracks. This is governed entirely by tape guides and head-drum ruler edge, and the angles and dimensions of these components are critical. After a period of use, dirt and grease deposits will build up on their surfaces, and that of the head drum. These can obstruct or divert the tape path, and the result will be mistracking. The first step in deck servicing, then, is to thoroughly clean the parts illustrated in the typical deck layout in *Figure 20.1*. They are the tape guides, rollers, ruler edge, head drum surface, capstan and pressure roller. If necessary the latter should be roughened with very fine emery cloth then degreased. At the same time, residual dirt, tape particles and grease should be removed from all deck surfaces to prevent its later migration on to critical components. A cotton bud and methylated spirit are the best tools here, the moistened bud forming an effective wiper and collector of debris. When all surfaces are clean and guides, rollers and capstan polished, head cleaning can begin.

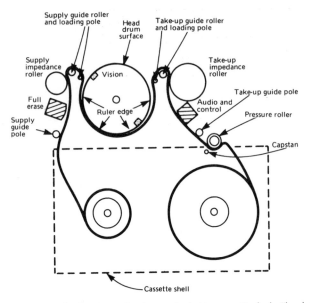

Figure 20.1. Cleaning points in a typical videocassette deck. The shaded parts are the tape heads, and cleaning of these should be left till last

For the static heads (full-width erase, sound and sync) a perfectly clean cotton or buckskin cloth should be used, moistened with alcohol; a cotton bud is useful where head surfaces are difficult of access, as in some Beta-format machines. Firm rubbing across the face of the heads is required, followed by final cleaning and polishing with a clean cloth or bud. The video heads are very fragile indeed, and cleaning should be undertaken with very great care. Manufacturers recommend the use of a lint-free (i.e. chamois or buckskin) cloth dipped in methanol, though many engineers prefer pure surgical spirit as it leaves less deposit. The cleaning motion should not be prolonged, and must always be made across the heads (i.e. in the direction of the tape path), *never* vertically. This is illustrated in *Figure 20.2.* Suitable cleaning materials and solvents are available from

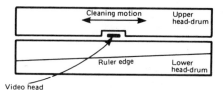

Figure 20.2. The cleaning motion for video heads must be side-to side. The video head is very delicate indeed; its width may be less than that of a human hair

VTR manufacturers. For inspection of heads and guides during cleaning a dental mirror is very useful, especially an illuminated type. After cleaning, it is important to allow all fluids to dry completely before tape loading, or the moist surfaces will 'pick up' the tape with dire results.

Tape guide alignment

There are several tape guides encountered by the tape on its path through the deck, as shown in *Figure 13.9*, and the most critical of these are the entry and exit guides at each end of the tape's wrap around the head drum. These determine the tape angle relative to the fixed video head sweep, and are made adjustable so that exact compliance with the format specification can be achieved. For correct setting-up, a standard is required, and this comes in the form of an *alignment* or *interchange* tape, pre-recorded at the factory on a precision standard machine under closely controlled conditions.

The entry guide determines head tracking angle at the point where the heads start their sweep of the tape, and examination of the RF *FM envelope* signal from the replay heads will indicate any fall-off in amplitude here due to guide misalignment. A good envelope signal is shown in *Figure 20.3(a)*, while *(b)* indicates low FM output at the LHS (start) of each envelope pattern due to incorrect entry-guide height. The exit guide is adjusted for optimum envelope shape at the

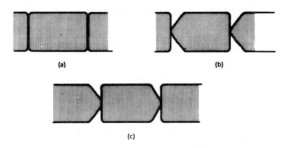

(a)

(b)

(c)

Figure 20.3. RF envelope patterns. (a) shows a good envelope, (b) one resulting from a dirty or incorrectly adjusted entry guide, and (c) the effect of dirt or maladjustment in the exit guide

end (RHS of oscillogram) of each head sweep, and (c) shows the effect of maladjustment here. Because of the FM limiting circuits employed on playback, a worst-case FM envelope shortfall of 40 per cent is tolerable before deterioration of the picture is noticed; this is illustrated in *Figure 20.4*.

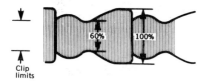

Clip
limits

60%   100%

Figure 20.4. The limiter will clip the FM luminance signal on replay, so that a shortfall of up to 40 per cent will pass unnoticed

The other tape guides are concerned mainly with defining the tape path across the static heads, and ensuring a parallel passage through the capstan/pressure roller combination. The adjustment of the sound/sync head is critical for height, azimuth and vertical tilt, and these are set up (on the audio signal from the alignment tape) by adjustment of the head's base mounting screws, in similar (but more critical) fashion to audio-recorder heads.

Tape tension is set by adjustment of the tension servo, in whatever form it takes (see Chapter 17), for a specific back-tension, monitored by a suitable gauge supplied by the manufacturer. Take-up tension is determined by the friction in the take-up clutch. This is either fixed during manufacture or is adjustable in three or four pre-set steps.

Reel drive, alignment and braking

The final aspect of deck servicing is concerned with the reel transport mechanics and operating levers. Their intricacy will depend on the vintage of the machine; early piano-key types require lubrication and occasional physical alignment of spindles, levers and slider bars; and roughening and de-greasing of rubber friction wheels and drive surfaces. Spirits can attack rubber compounds, and a mild detergent solution

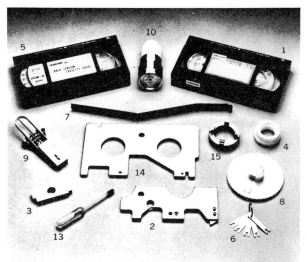

Figure 20.5. Tools, jigs and test cassettes for use in VTR deck servicing: (1) MH2 alignment tape. (2) Master plane jig. (3) Height gauge. (4) Master plane position jig. (5) Back-tension gauge. (6) Thickness gauges (set of five). (7) Cassette holder. (8) Tension-check dummy reel. (9) Tension gauge. (10) Torque gauge (with pedestal). (13) Oil (capsule). (14) Cassette housing jig. (15) Frequency gear PWB jig (Ferguson Ltd)

is recommended for cleaning and degreasing drive belts and rubber wheels. Sometimes this is easier said than done, and replacement rather than refurbishing of interwheels, brake pads and drive belts is often the most practical course. Later machines with direct-drive systems are easier to maintain in this respect.

There are several other critical adjustment points in the video tape deck, such as reel disc height, tension pole positioning and video head eccentricity, and the setting of these is comprehensively covered in the relevant service manual for each model. Jigs and tools are available from VTR manufacturers to assist with precision adjustment, and a set of these for one particular machine is pictured in *Figure 20.5*.

Electrical adjustment

Many of the cautions mentioned above are also relevant to the electrical pre-set controls within the machine. Although electrical setting-up (especially of those controls which concern servo and video-head response characteristics) is a part of routine maintenance, electronic devices are not subject to wear; and little drift is encountered in modern circuits, especially the burgeoning digital sections. Electrical adjustments tend to be specific to each model and format, so that little general guidance can be given here, except to say that service manuals give precise setting-up instructions to be used with the necessary test gear.

Test equipment

For VTR circuit investigation and servicing, it is important that the test equipment used be adequate, accurate and dependable. An oscilloscope of minimum specification would be 10 MHz bandwidth, 2 mV sensitivity and dual-trace capability. High-impedance multimeters of both digital and analogue type are required, the former with a DC voltage accuracy of 1 per cent or better. For investigation of colour-under and servo electronics, a frequency counter is essential, preferably a type with 6½ or more digits and an accurate timebase.

Necessary signal sources are a bench-type colour bar signal generator with split-field display and outputs at video (baseband) and UHF; an accurate signal generator with sine- and square-wave outputs; and a source of noise-free off-air transmission signals at UHF. For monitoring purposes a modern high-performance TV receiver with rotary tuning is ideal, preferably with a video input in monitor fashion. A mains variac is very useful for investigating power-supply and excessive-loading faults; a mains isolating transformer is also important for safety, especially when dealing with the sort of switch-mode mains power supply described in Chapter 19.

In addition to these, alignment tapes and adjustment jigs are required as described earlier in this chapter, along with a full service manual for the machine under investigation. The final factor is a good working knowledge of the principles and practice of the VTR!

Output channel and still-frame adjustment

Some VTR models have rear-mounted pre-set adjustments for use during installation, and the two most common will be described. When the VTR is being used in the environment of other RF generators around TV channel 36, the machine's RF output will beat with them, giving rise to dots or a wavy or herringbone pattern on the reproduced picture. Although no broadcast TV transmitters operate in this region, some radar installations give rise to spurious outputs at about 600 MHz which can upset VTR reproduction. Increasingly, other UHF signal sources are finding their way into the home, in the form of TV games, computers and Teletext adaptors, all with UHF modulators operating in the region of channel 36. To avoid clashing with these, the VTR's UHF modulator can be varied over (typically) eight channels by adjustment of a trimmer in the RF modulator. When adjusting this, go only so far as is necessary to eliminate beat effects; RF output level drops off at the extremes of adjustment, and on a more practical plane, the tuning slug may drop into the works if it's screwed in too far!

Where noise-free still-frame facilities are provided in a VTR, there may be an external adjustment for pre-setting the 'artificial' field sync pulse timing during this mode, as described in Chapter 17. In 'freeze' mode a degree of judder may occur if successive fields differ greatly in content, as would be the case when televising a fast-moving athlete. The ideal frozen frame for this adjustment, then, is a test card or similar stationary picture with well-defined horizontal edges, on which the control should be set to 'null-out' vertical judder in the displayed still frame.

Television time base requirements and modifications

In our discussion of luminance and chrominance playback techniques (Chapter 14), we came across the phenomenon of timing errors in the off-tape replay signal, known as jitter. The deck maintenance and lubrication procedures just described go a long way to minimise jitter, but it is always present in the VTR's output signal and can upset the horizontal synchronisation of the TV receiver (or monitor) unless the latter is designed for use with a VTR. TV sets are primarily intended for broadcast reception, and in noisy or fringe reception areas line synchronisation can be upset by interference and electrical noise. To overcome this problem, they are fitted with a flywheel line-synchronisation circuit in which the timing of the line sync pulses over several lines' duration is 'averaged' to effectively slow down the response of the line oscillator to them. Thus occasional ignition interference or noise pulses occurring just before the end of a TV line are prevented from falsely triggering the scan generator and giving a ragged appearance to the picture.

If the flywheel time constant is too long, the timing jitter on the VTR output signal (which is present, of course, on picture information as well as line syncs) will cause the picture to twitch horizontally while the timebase produces a smooth, average-rate scan timing. This effect is particularly annoying, and to prevent it the flywheel time constant must be reduced to enable the timebase to follow the rapid off-tape timing fluctuations, thus straightening up the picture. For many years TV design has taken account of this, and a dedicated VTR button on the set automatically switches in a suitable time constant in the flywheel department in sets where this is not present anyway. Early sets (and even some of quite recent vintage) need modification to their flywheel circuits to fit them for good video playback, and this generally involves reducing the time-constant to about one-third of its original level – specific modification details are available from set-makers and VTR manufacturers.

Receiver-monitors

Many contemporary TV sets have advanced power supply circuits along the lines of *Figure 19.5* which give mains-isolation to the 'ground' line of the set, and this paves the way to the provision of input sockets for audio and video signals at baseband. The TV becomes known as a receiver/monitor, and where this type of set is available, much better replay performance in vision and sound is possible by direct connection to the VTR. Baseband linking eliminates the degenerative processes of converting to and from UHF and IF, and has much to recommend it. The use of a purpose-designed precision monitor is not really justified with domestic VTR machines, as system performance is limited by the VTR capabilities, which fall far short of those of the monitor.

Distribution of VTR output signals

Where a VTR is required to operate several displays simultaneously, perhaps at widely different points, there are two choices in the mode of distribution. For optimum quality and avoidance of spurious effects due to patterning and timing errors, the ideal is to distribute at baseband. This involves a coaxial cable carrying composite video at 1 V pk/pk to monitors or monitor/receivers. A video distribution amplifier is needed to provide outputs at 75 Ω impedance for as many monitors as required, while the sound signal is either carried in a separate cable or sent along the video coax, with insertion and extraction by suitable low-pass filters at each end. Such a system is ideal for teaching, conferences and similar applications and will operate at distances up to 700 metres without serious distortion.

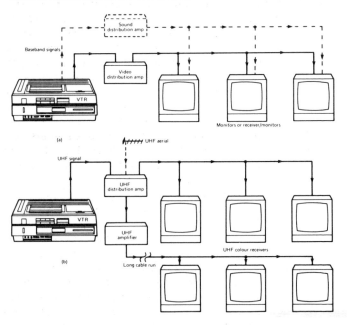

Figure 20.6. Two forms of signal distribution for VTR playback. That given at (a) is capable of best results

The alternative is distribution by an RF carrier, and so far as domestic VTRs are concerned, this implies the use of the UHF output from the machine. If more than two monitoring points are needed, a UHF distribution amplifier will be necessary, and by this means many sets can be fed with the playback signal, at great distances if required. This method, while offering lower picture and sound quality than the baseband system, has the advantage of convenience in the use of easily-available equipment. The two systems are shown in *Figure 20.6a* and *b*.

Dubbing and editing

As VTR design grows up, new features are being added to the basic machine, and the facility to assemble, dub and edit programmes represents, along with the availability of reasonably priced colour cameras, a significant incursion into the 'home-movie' market. Audio dubbing is carried out by operating the machine in replay so far as the vision signals are concerned, but with the sound section of the machine in record mode, and the audio input jacks 'live'. Apart from the necessary electronic switching facilities, the machine needs only an additional audio erase head (positioned on the tape path just before the audio rec/play head) to achieve this, and in dub mode the replay picture is monitored while adding a new sound track to the recording.

Video editing still suggests to many people some sort of surgical operation on the tape itself, and while this is commonplace in professional tape systems, any attempt to cut or splice cassette videotape in domestic environments spells doom for the machine's video heads! Editing, then, consists of stopping and starting the VTR's transport system to 'cut' between required programme sequences. Two forms of editing are possible, known as *assembly edit* and *insert edit*. Assembly edit is used during recording, and in effect stops the recording process while the next sequence is made ready; it can be used with any video source(s), but most commonly its use is with a video camera via its pause, or *run*, button. Insert editing is the 'dropping-in' of a sequence into a pre-recorded programme, examples being titles, credits or commercials.

While these processes can be carried out on any machine by the use of record and pause controls, the edit point will be marred by spurious effects. There are a number of reasons for this. During record and playback the servos, as we have seen, are locked up to incoming field syncs and off-tape control tracks respectively. When a change is made in input signal the servos have to run up to speed and lock in to the new video source and so the signal recorded on the tape goes haywire during this period. On replay the discontinuity in the control track also causes a 'hiccup' in servo operation, and these effects result in a very ragged transition between the two pre-recorded sequences. Further problems arise from the fact that the full-width erase head is placed about 8 cm before the video head drum on the tape path, giving rise to three or four seconds crosstalk from the original tape tracks before newly-erased tape reaches the recording heads.

'Clever' edit

To overcome these problems and give a reasonably clean transition at edit points, the technique of *back-space edit* was introduced. In its most common form it involves the VTR machine back-spacing (or rewinding) 20 to 25 frames (about a second) when record pause is selected, then stopping. At the end of the pause period the machine will re-start in playback

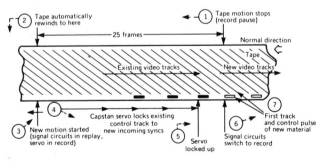

Figure 20.7. Assemble edit: The sequence of events is indicated by the ringed numbers, and the result is a clean switch from old to new programme material

mode so far as its vision systems are concerned, but with the capstan servo slaved to the new incoming video signal. When the servo has locked up, the vision circuits switch automatically to record mode during a convenient field-blanking interval so that the last frame of the old material is followed by a new and fully synchronised field of the new programme with no lack of continuity in control track or video tape tracks; very little disturbance will be seen on the replayed programme over the edit point. A certain amount of chroma crosstalk can occur briefly at the changeover point in some designs, and much depends on the sophistication of the VTR's electronics – this is another area in which the microprocessor has come to the fore. An idea of the 'clever-edit' scheme is shown in *Figure 20.7*.

Portable VTRs

The earliest portable VTR equipment for domestic use consisted of separate camera and shoulder-hung recorder; current portable VTRs incorporate the camera on-board.

Although portable VTRs are carefully and specifically designed for their purpose, especially in the areas of servos and power consumption, some limitations exist quite apart from photographic and aesthetic ones. Let's briefly examine two of them. In use, a portable VTR will necessarily be moved about; in some situations like motor rallies and fairgrounds, sometimes quite violently! This will have a gyroscopic effect on the rotation of the operating shafts in the machine, and particularly the relatively heavy head drum. To overcome this, drive motors in such machines have multi-pole windings, and a special HF tacho generator in the servo circuit ensures that speed correction takes place very quickly. In spite of these, sudden and violent movement of the machine will upset recordings and so should be avoided if possible.

On location both VTR and camera sections depend on battery power. Where the system can be hooked up to a 12 V car or boat supply this is no problem, though incorrect polarity of connection will bring the shooting expedition to an unhappy and premature

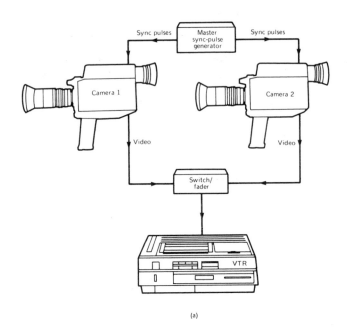

end! More often a Ni-Cad battery, clipped onto the VTR, is used and this will give between 30 and 60 minutes recording time, depending on VTR design. Time passes very quickly under these circumstances, and so Ni-Cad batteries can prove something of a stumbling block. Several fully-charged ones are really needed before setting out, and it is difficult to establish how much charge is present in a unit before use. Over-charging will quickly destroy the batteries, and excessive discharging is not good for them either. To avoid frustration and ensure a reasonable working life from these batteries, it is necessary to follow their manufacturer's instructions to the letter.

Synchronisation of signal sources

In our discussion of editing we touched on the making of videotape programmes with the use of a TV camera, and

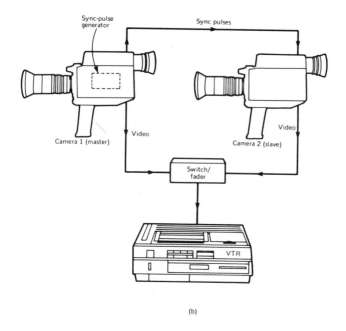

(b)

Figure 20.8. Alternative methods of slaving TV cameras or other video sources. (a) sync pulse generator and two slaves; (b) master-slave system

many other signal sources are available, both optical and electronic. Slide scanners, telecines, pattern and caption/character generators are becoming available ·to amateurs as well as semi-professionals, and where two signal sources are required to be used in the making of a programme they need to be scan-synchronised to each other. This avoids disturbance when switching between them, and is essential if fading or mixing operations are to be carried out. Two 'slaving' arrangements for this purpose are shown in *Figure 20.8*. Regrettably, manufacturers have neglected this aspect of programme-making in the past, and many cameras and other video sources do not have slaving facilities.

405

VTR operation on non-standard signals

There is some confusion over the question of inter-changeability of VTR machines and tapes between different countries of the world, probably arising from the fact that all three major formats are used world-wide, and it's reasonable to suppose that, for instance, a Beta-format colour tape recording made in the USA will replay on a UK-type Beta machine. In fact all three basic formats are manufactured in export versions for different world markets, but within each format some differences are present in the electronics to cater for variations of colour encoding (PAL, SECAM, NTSC), line frequency (525 or 625 lines) and field rate (50 or 60Hz), and these determine the circuit design of servos, colour-under systems and still-frame facilities.

The different transmission standards between countries add further complication, this time in the peripheral circuits of the VTR such as tuners, sound detectors and RF modulators. Vision and sound modulation systems differ, and VHF bands are used for TV transmission in many parts of the world. As a result, UK-type machines and tapes will only work in Eire, Hong Kong and South Africa, and even then an adjustment to the mains voltage tapping is required! We shall see in the next chapter the effects of incompatibility between formats rather than TV standards.

Some manufacturers market multi-standard machines (currently available in VHS and Betamax formats) which go some way towards solving the compatibility problems arising out of differing TV systems. No such thing as a truly univeral machine exists, however, and if a multi-standard machine is considered for purchase, it's important to study the specification carefully and establish that it will do what is required of it; these machines are usually only available to special order.

21

Tape formats — systems and facilities compared

In preceding chapters we have examined VTR principles and circuits in general terms, taking examples from all home formats to illustrate the techniques used. Where differences between formats have arisen in the text they have been surprisingly few, and mainly confined to Video-8. In this last chapter we shall expand on the format differences and look into their significance, both from the point of view of the consumer and those concerned with the technicalities of the machines.

There are currently three major home formats in the consumer field: Betamax, Video-8 and VHS, the latter having two very distinct variations. Some contenders in the format wars are illustrated in *Figure 21.1*. Betamax and VHS owe their origins to the Japanese, Sony and JVC respectively; they are very similar in almost all respects except the vital one of compatibility with each other! The Video-8 format is a later development, using advanced tape and features — we have met ATF, AFM and PCM sound in previous chapters. Judged subjectively on normal programme material, the performance of the three competing formats is quite similar in terms of picture resolution and signal/noise ratio.

The existence, side by side, of several incompatible formats matters not a jot when timeshift recording is required, but comes very much to the fore when any form of videotape interchange is contemplated. Examples are many: hiring of cassettes for private entertainment or instruction; distribution of cassettes by educational bodies to schools and colleges; exchange of letter-tapes between distant relatives and friends; and storage of archive material which may not be required for many years.

(a)

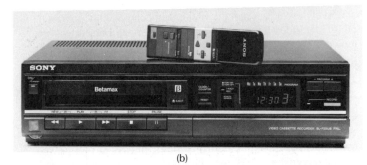

(b)

Figure 21.1. The contenders: (a) standard VHS: JVC model HRD53OEK; (b) Betamax; Sony model SL-F25UB; (c) Video-8: Sanyo camcorder model VMD5P; (d) VHS-C: Bauer-Bosch camcorder model VCC616AF

408

(c)

(d)

Apart from the consumer's dilemma as to which format to purchase, the main burden is carried by the middlemen of the industry, such as tape libraries and hire-shops, who must duplicate or triplicate their stock of software with little financial reward; video hardware dealers, who must stock a wide range of equipment in various formats and sub-formats; and the VTR service industry, whose hard-pressed technicians need to have test equipment, jigs, documentation and in-depth knowledge of all current and obsolescent formats and their workings. Many years ago standardisation of *sound* recording systems was achieved at the outset, with 12.7 mm ($\frac{1}{4}$ in) tape and a useful choice of speeds for open-reel systems, and then the brilliantly successful and convenient Compact Cassette system for home use. The latter, as a worldwide industry standard, has reached a high degree of development, and is only now starting to be challenged by the R-DAT tape system.

In the VTR world, availability of pre-recorded and blank videocassettes counts for a great deal, and history has proved that the earliest and best-established formats survive and prosper at the expense of newcomers, particularly those which offer no *perceptible* advantage in performance, features, price or convenience. This accounts for the present dominance of VHS, which has ousted V2000 format, and thoroughly beaten Betamax in terms of units sold. The more successful challengers to VHS owe their existence to the definite advantages they offer: Video-8 for its excellent sound, small light cassettes and high performance in a mobile (camcorder) role; and S-VHS for its excellent picture resolution and clarity, even though the actual recordings are not compatible with standard machines.

The existence of three-and-a-half incompatible formats – and two cassette-size variations for VHS camcorders – is inconvenient for the user and sometimes exasperating for the trade, but is indicative of the atmosphere of free enterprise and unrestricted marketing which we enjoy in the free world. The spur of competition, as in other spheres, confers a seldom-realised advantage for the consumer in that he/she is able to purchase and enjoy a very advanced and sophisticated VTR, bristling with features, at an artificially low price. The situation has arisen because manufacturers and retailers have played leapfrog

410

amongst themselves, in their respective spheres of features and profit margins, in an effort to promote themselves in the marketplace. Manufacturers have the long-term objective of building their reputations, pet formats and market-share; while retailers, locked in the stranglehold of high street competition, have the simple objective of staying in business by maintaining a high volume of low-profit turnover.

Let us now examine the main features of each format.

Betamax

This system was designed by Sony of Japan, and introduced to Europe in 1978. Initially Betamax machines sold well and enjoyed a significant market share. In recent times, however, the popularity of Beta has waned for software-availability reasons.

The head drum in a Beta machine is 745 mm in diameter, significantly larger than that of all the other formats. A large head-drum means a high peripheral (writing/reading) speed, and with all other factors being equal it was held by many that Betamax pictures were better than contemporary rival formats.

Betamax cassettes are smaller than standard VHS types, and fit easily into a jacket pocket. This was one of the earliest design objectives, and the cassette is light and convenient as shown in the comparative illustration of *Figure 21.2*. Some of the tape packages are shown in *Figure 21.3*, with (top) an audio cassette for comparison purposes. Although not part of the format specification, Betamax machines invariably use a *U-wrap* for tape threading, as shown in *Figure 13.7*. Later Betamax machines have a very small threading ring, which closely surrounds the head-drum.

Electrical characteristics of Beta VTRs are similar to those of VHS. The same type of tape formulation is used, and the conception and birth periods of these two formats coincided. The Beta chrominance system was described in Chapter 15, and few other differences exist. Although not part of the format specification, Beta machines were the first to have a circuit incorporated to reduce crosstalk in luminance signals, a practice which has now become widespread.

We have seen that crosstalk effects between recorded tracks are

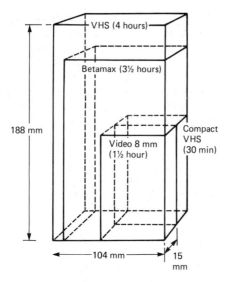

188 mm

104 mm | 15 mm

Figure 21.2. Cassette sizes and playing times compared. The two smallest cassettes shown (V8 and VHS-C) are the most common candidates for LP operation, giving 3 hour and 1 hour running time respectively

most troublesome at low frequencies, and as *Figure 15.16* shows, the lower luminance frequencies are vulnerable to crosstalk effects, giving rise to a fine patterning and 'orange-peel effect' on the TV screen. To eliminate this, the recorded luminance FM carrier is frequency-shifted by 7.8 kHz (half of line frequency) between adjacent video tracks during record, so that on replay a 180° phase difference can be generated between wanted and crosstalk signals. This technique is familiar from our study of chrominance recording in Chapter 15. Here the luminance crosstalk is removed by a one-line delay and add-matrix, just as in the colour crosstalk removal filter illustrated in *Figure 15.19*.

VHS

VHS stands for Video Home System: this format was designed by the JVC company in Japan. Like Beta it was released in Europe in

Figure 21.3 Cassette size comparison: from the bottom, VHS, Betamax and Video-8. At the top is shown an ordinary audio cassette for comparison purposes

413

1978. VHS, however, enjoyed the backing of one of the leading UK rental companies of those days, Thorn-EMI—Ferguson, and this largely accounts for its fast and deep initial penetration into the UK market, aided by the British habit of renting rather than buying television equipment. Simple mechanics characterise the VHS scheme, with an M-wrap threading system using combined threading posts/tape guides as in *Figure 13.6*. A tilted head drum also simplifies the deck system by permitting a tape path which is parallel to the plane of the deck service, a layout now adopted in all formats.

Standard VHS boasts the widest video track of any current format (49 microns) which offers best signal/noise ratio conditions for the video signal. The small size and physical simplicity of VHS deck and tape transport arrangements (especially the small-head variations described in Chapter 13) lend themselves well to portable machine applications, and it was in VHS form that the first 'mobile' VTRs appeared. Indeed in the UK innovations have tended to appear first in VHS machines, and this is true of many of the features described earlier in this book, such as still-frame, trick-speed, clever editing and stereo sound.

A clever adaptation of the standard tape and cassette package is VHS-C (Compact) which permits a small and light VHS camcorder. The cassette is a small (92 × 59 × 23 mm) housing containing about 30 minutes worth of standard 12.7 mm VHS tape. It fits a very small camcorder (*Figure 21.1d*) weighing less than 2kg, and incorporating a small head drum, thin direct-drive motors and a solid-state image sensor. Back at home the small cassette is loaded, piggy-back style, into a normal-size adaptor shell for replay or editing in the standard VHS machine.

VHS-LP

Both camcorders and homebase machines are available with a dual-speed option. On record they are switchable between standard and slow speed, and during replay can automatically recognise which recording speed was used in any cassette offered to them.

414

In LP mode the capstan speed is halved – to double the playing time of a standard cassette to a maximum (E240 tape) of eight hours. Early dual-speed machines had separate head pairs for SP and LP modes; later designs incorporated both SP and LP heads in a single pair of ferrite chips; current practice is the use of the *same* pair of heads in both modes. Here the head width is a compromise; in SP, narrower-than-standard tracks are written, leaving between them a form of guard-band. In LP, each recorded track is wider than required, and its excess width is cut down by the erasing action of the next head sweep – see *Figure 13.21* and associated text. LP video tracks are 24.5 microns wide, and the sound and control tracks are recorded in the normal way at the slower speed. The track patterns, then, represent a 'telescoped-up' version of *Figure 13.5* with video tracks and control pulses now occupying half the linear tape space of those of a standard VHS recording.

Performance-wise, VHS-SP is surprisingly good. Video S/N ratio is slightly impaired due to the narrow video tracks; but sound HF response is dramatically cut, where longitudinal tracks are used. Many non-critical viewers cannot tell the difference between SP and LP on replay, so long as it comes from a high-performance tape in good condition. The maximum eight-hour capability of VHS-LP comes into its own when used in conjunction with the multi-event timers now commonplace, and allows, for instance, eight separate one-hour recordings to be made without human intervention.

Even though LP performance is good, its use in camcorders and homebase machines is not recommended unless the programme will be viewed once (non-critically!) before being discarded.

S–VHS

Super-VHS is an advanced new variant of the established format. It uses a high FM carrier, with deviation (*Figure 21.4*) from 5.4 to 7 MHz and more 'meat' in the video sidebands. To permit this an advanced new videotape formula is used in a cassette of conventional size, shape and running time. A very small video head-gap, and 'fine-grain' magnetic tape permits a baseband video frequency

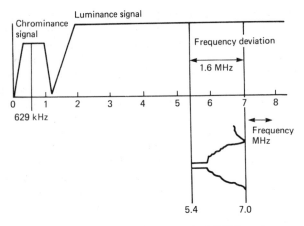

Figure 21.4. Recording signal spectrum for S-VHS format

response approaching 5MHz, and on-screen resolution better than 400 lines. Signal/noise ratio is also better than that of the other formats.

One of the problems of PAL and NTSC colour encoding is the presence of cross-colour effects, discussed in some detail at the end of Chapter 4. Where the programme is originated in these forms little can be done to prevent cross-colour – it stems from the decoding process at the receiving end. For other applications (home videography, MAC transmissions, pre-recorded movies on tape, etc.) the S-VHS system has provision for complete separation of Y and C (luminance and chrominance) signals throughout the record and playback processes. Where the TV receiver or monitor is equipped with a suitable input socket (*S-terminal*) and a large screen, S-VHS format is capable of a picture performance superior to that available from a terrestrial broadcast system.

Video-8

The Video-8 format uses 8 mm-wide tape, from which it gets its name. Unlike the others this system was developed by a consortium of companies and accepted for use by over 127 of the world's major audio/video manufacturers. It is the first domestic video system to use digital audio recording, and the first to be designed from the outset for an alternative mode of operation – as a high-quality digital sound-only recording system. Its other advantages are a tiny cassette (*Figure 21.3*), facilitating miniaturisation of the equipment, be it portable or homebase type; a flying erase head for good edits, a feature now appearing in some VHS equipment; and the exploitation of new tape and head materials and techniques for better performance.

Although offered in homebase form for tabletop use, Video-8 is seen mainly as a camcorder format, whose primary advantages are excellent sound, light weight, and high performance – the Video-8 manufacturing companies are particular experts with lenses and TV image sensors.

The Video-8 tape wrap was shown in *Figure 16.8*. Although the format specifications do not in any way dictate the mechanical layout of the machine, V8 machines commonly use a combination of M-wrap and U-wrap techniques as shown in *Figure 21.5*. At *a* is shown the situation where the cassette has just been lowered onto the deck: the pinch-roller and several tape guides have penetrated the cassette behind the front tape loop. In diagram *b* the first (M-loading) phase is complete: two guides have drawn a loop of tape away from the cassette, VHS-fashion. At this point the loading ring starts to rotate anticlockwise, further wrapping the tape around the head drum, borne on a guide/post

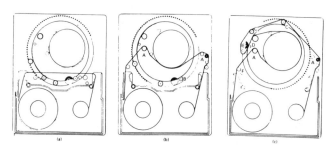

Figure 21.5.   Video-8 tape threading: (a) cassette in; (b) first stage; (c) wrap completed. Compare with *Figures 13.6 and 13.7*

ahead of the pinch roller, and prevented from folding back on itself by further ring-mounted guides, diagram c.

Video-8 tapes have a maximum playing time of ninety minutes in standard-play, doubled to three hours in LP mode, for which all V8 machines are equipped.

Format comparison

The salient features of the competing formats are shown in *Table 21.1,* which includes details of some obsolete formats for comparison purposes. The VCR, VCR-LP and V2000 formats are really only of academic interest now, but are included to show the progress made over the years.

Regarding performance, too much emphasis should not be put on the details shown in Table 21.1, since it is very difficult, even for those experienced in domestic VCR use, to tell the formats apart by watching replayed pictures from any of them - apart from S-VHS, whose pictures are markedly sharper than those from all other formats. In average domestic conditions, any slight performance differences between the various systems are sometimes 'swamped' by other shortcomings; a chain is only as good as its weakest link, and many factors beyond the control of the VTR manufacturer can mar performance. A weak aerial signal impairs signal/noise ratio; worn tape or dirty heads cause excessive drop-out; maladjustment of the tracking control, or a deck in need of servicing leads to mistracking. The TV set in use may not be in perfect condition, and maladjustment of the focus control, a worn tube or incorrect tuning can easily halve the definition available from the VTR!

Special features

The provision of extra features has a value which varies greatly from user to user, and in some cases the advantages of these are more apparent than real. All VTRs are capable of recording and replaying TV programmes well, and for the reasons just given, and the fact that signal-processing circuit design does not vary much

Table 21.1 Formats compared: the main parameters of past and present home formats. All but Video-8 use 12.7 mm ($\frac{1}{2}$ in) tape

	Linear tape speed, cm/sec	Video writing speed, m/sec	Video track width, microns	Azimuth offset of video heads	Head drum diameter, cm	Video track angle to tape	Sound track width, mm (mono or L + R)	Max playing time, hours	Information density on tape (hours per sq. metre)
VCR	14.29	8.10	130	± 15°	10.5	3.69°	0.7	1	0.16
VCR-LP	6.56	8.10	85	± 15°	10.5	3.71°	0.7	2	0.33
VHS	2.34	4.85	49	± 6°	6.2	5.33°	1.0	4	0.93
Beta	1.87	5.83	33	± 7°	7.45	5.97°	1.05	$3\frac{1}{2}$	1.16
V2000	2.44	5.08	23	± 15°	6.5	2.65°	0.65	8*	1.79
Video 8	2.01	3.12	34	± 10°	4.0	4.92°	N/A	$1\frac{1}{2}$*	1.70

*Flip-over cassette; 2 × 4 hours
+ LP mode: 3 hours

419

between machines (leaving out the special case of S-VHS format), a £700 model is unlikely to give markedly better performance on normal programme replay than one costing half as much. Most domestic VTR use is for recording and replay of films, TV serials and sports events and for these a basic machine is quite adequate.

Most of the extra costs of full-feature VTRs, then, go into the provision of still frame, remote control, timer programming, stereo sound, and trick-picture effects using field-store memories, with the basic electronics and mechanics largely unchanged. In many households the use made of many of these facilities does not justify their cost. Certainly remote control (almost universal now) is worthwhile, and good freeze-frame and frame-advance facilities are very useful in educational spheres, and for sports enthusiasts. The timer versatility needed depends very much on the user's time commitments and viewing tastes, and a one- to three-programme capability over one week will cater for the requirements of a goodly percentage of the population. A great deal of money can be saved by careful thought before buying VTR equipment!

Index